Palgrave Studies in the History of Science and Technology

Designed to bridge the gap between the history of science and the history of technology, this series publishes the best new work by promising and accomplished authors in both areas. In particular, it offers historical perspectives on issues of current and ongoing concern, provides international and global perspectives on scientific issues, and encourages productive communication between historians and practicing scientists.

More information about this series at
http://www.palgrave.com/gp/series/14581

Samuel A. Robinson

Ocean Science and the British Cold War State

palgrave
macmillan

Samuel A. Robinson
Department of Sociology
University of York
York, UK

Palgrave Studies in the History of Science and Technology
ISBN 978-3-030-10314-9 ISBN 978-3-319-73096-7 (eBook)
https://doi.org/10.1007/978-3-319-73096-7

Printed on acid-free paper

This Palgrave Macmillan imprint is published by the registered company Springer International Publishing AG part of Springer Nature.
The registered company address is: Gewerbestrasse 11, 6330 Cham, Switzerland

For Charmaine

Acknowledgements

Research for this book was made possible by the project "The Earth Under Surveillance" funded by the European Research Council with grant no. 241009. It started life as a PhD dissertation at the Centre for the History of Science, Technology, and Medicine, University of Manchester, under the supervision of Simone Turchetti and Jeff Hughes. It was revised into its current form while I was a postdoctoral researcher on the AHRC-funded project, "Unsettling Scientific Stories" in SATSU, Department of Sociology, University of York, where PI Amanda Rees provided endless support, and my fellow postdocs Matthew Paskins and Amy Chambers kindly offered comments and criticism on chapter drafts, significantly improving the readability of this manuscript. Any remaining errors naturally remain entirely mine.

I am grateful to the librarians and archivists of the National Oceanography Centre Library, Southampton; the National Archives, Kew; John Rylands Library, University of Manchester; Churchill College Archive Centre, University of Cambridge; the British Library; and the Caird Library, National Maritime Museum, Greenwich, whose expertise and enthusiasm for their collections made the process of researching this work a pleasure.

To Amy, Rupert, Rosalind, and Rufus, you have got me through the months and now years when I thought this project would never end. I am grateful and look forward to our next book adventure.

Finally, this whole process would never have been possible without the love and support of my mother, Charmaine. Ever since I demanded history books for my seventh birthday, history and my ever-expanding library have consumed your house and probably your life. Thank you for almost never complaining. You have taught me how to be, and for that I will forever be in your debt. This book is dedicated to you.

Contents

List of Abbreviations

ACSP	Advisory Committee for Science Policy
AFMED	Allied Forces Mediterranean (NATO)
ARE	Admiralty Research Establishment
ARL	Admiralty Research Laboratory
ASDIC	Allied Submarine Detection Investigations Committee
ASW	Anti-submarine warfare
AUWE	Admiralty Underwater Weapons Establishment
BAAS	British Association for the Advancement of Science
CRNSS	Chief of the Royal Navy Scientific Service
DPR	Director of Physical Research (Admiralty)
DSI	Director of Scientific Intelligence (Admiralty)
DSR	Director of Scientific Research
ELINT	Electronic intelligence
FRS	Fellow of the Royal Society
GIUK	Greenland–Iceland–UK gap
ICES	International Council for the Exploration of the Seas
IGY	International Geophysical Year
IOS	Institute of Oceanographical Sciences (rebrand of NIO)
IUGG	International Union of Geodesy and Geophysics
JIB	Joint Intelligence Bureau
JIC	Joint Intelligence Committee
JOC	Joint Oceanographic Commission
MAFF	Ministry of Agriculture, Fisheries and Food
MILOC	Military Oceanography Group, NATO
MMD	Mine Design Department, Royal Navy

MOD	Ministry of Defence
NATO	North Atlantic Treaty Organisation
NERC	Natural Environment Research Council
NIO	National Institute of Oceanography (UK)
NOC	National Oceanographic Council
OSINT	Open-source intelligence
RAF	Royal Air Force
RNSS	Royal Navy Scientific Service
RRS	Royal Research Ship
SAC	Scientific Advisory Committee
SERL	Services Electronic Research Laboratory
SIGINT	Signals intelligence
SOSUS	sound surveillance system
SSK	Sociology of Scientific Knowledge
TRE	Telecommunications Research Establishment
UNCLOS	UN Convention on the Law of the Sea
WHOI	Woods Hole Oceanographic Institute

List of Figures

CHAPTER 1

Ocean Science and the British Cold War State

In March 1974, a one-day symposium was held at the Royal Society in London, to reflect on the effect of two world wars on the organisation and development of science in the United Kingdom.[1] Listed amongst the speakers were two leading British scientists who feature prominently in this book: Edward Bullard and George Deacon. Bullard believed that important lessons taught to young scientists by the war included:

> how to use the Government machine, how to get one's way with committees, how to persuade people with arguments suitable to their backgrounds and prejudices and how realistically to assess the means needed for a given end.[2]

Following Bullard's paper, George Deacon (the recently retired director of the National Institute of Oceanography, NIO) noted that 'Marine science was revolutionised by the two world wars, particularly the second'.[3] In fact, their own post-war work took courses opposite to those that their statements at the symposium may indicate. Bullard had devoted more time to research than patronage, contributing to the establishment of the Cambridge-based Department of Geodesy and Geophysics, whereas Deacon fitted Bullard's description of the scientist's changed role, having mastered the art of "getting one's way" and using the government machine to drive new marine science.

S. A. Robinson, *Ocean Science and the British Cold War State*,
Palgrave Studies in the History of Science and Technology,
https://doi.org/10.1007/978-3-319-73096-7_1

Bullard's candid description of the interface between science and government, with its depiction of the scientist as the manipulator of the machine, is compelling. The extent to which his statement reflects the reality of post-war British oceanography is the central research objective of this book, which focuses on the relationship between ocean scientists, military officers, and government officials in Britain. These groups came to work together as a result of an increasing perception within the military that the ocean environment was dangerously unknown, affecting potential military readiness in a time of Cold War tensions.

Ocean Science and the British Cold War State seeks to address two vital questions: what kind of relationships existed (and developed over time) between ocean scientists, military officers, and government officials? And how does the study of these relationships contribute to our understanding of the development of Cold War science, especially in terms of patronage, policy, and resources? In order to address these questions, the book examines individuals involved in these relationships and their actions, in an effort to break down the monolithic treatment of scientific institutions, political departments, and the military.[4] This is a study of Big Science[5] as an instrument, with a focus on the individuals who played a key role in the political, military, and scientific networks in Britain that shaped the disciplinary trajectory of oceanography.[6]

The focus is primarily upon the actors within these large networks. I employ the notion of a "bio-network" to explain how certain individuals in a scientific community have the power to change the development of policy networks that affect the trajectory of their scientific discipline. Of course not all the historical actors in the book can be considered "network brokers". All of them make various entrances and exits; some play fleeting roles whilst others are present throughout, and some are versatile players who act different parts at different stages. One, however, has a recurring role: George Deacon. Yet this is not a biography of a leading British oceanographer of the period. It is concerned with the network that he worked to forge, maintain, and adapt in the context of British military, political, and scientific circles.[7]

Biographies of Hybrids in the History of Science

If this is not a biography, why discuss biography at all? For a long time, the history of science was seen as the account of Great Men through time.[8] As a result, biographical approaches within the field are often seen as tainted by a historiographical debt to hagiographical writing.[9] This hagiographical-biographical method was challenged after the Second World War by Robert Merton, who used his sociological approach to science studies to respond to a perceived need for science in the 1950s to promote a bigger, more collaborative vision of itself in light of the contemporaneous development of Big Science.[10] Merton's approach was the foundation of the sociology of scientific knowledge (SSK), which is concerned with understanding how 'types of social arrangements…were conducive to the production of certified knowledge'.[11] Essentially, Merton argued that the scientific community was central to the reception of scientific facts and theories, pointing out that the Great Men of science were often challenged and in their own lifetimes their theories were rarely accepted because of any ingrained notion of their "genius". Later scholars further refined these terms of reference to study, exclusively, the social construction of knowledge within science, in doing so discrediting biographical accounts.[12]

This approach typified ground-breaking works such as Steven Shapin and Simon Schaffer's study of the debate between Hobbes and Boyle over the latter's air pump experiments, and it can also be traced in Andrew Pickering's work on twentieth-century particle physics.[13] As scientific debate naturally involved more than one actor, SSK scholars championed the study of science as a community discipline and knowledge as something that could never be defined by the single individual alone, thus challenging the paradigm of the lone genius. In broadening the scope of study to scientific communities, Shapin and Schaffer demonstrated that natural knowledge in seventeenth-century Britain was shaped by contemporary political philosophies. The more obvious conclusions of the SSK movement, namely that scientists do not work in isolation and that they are situated within cultural milieus that extend beyond science, are very important to this study, which shares with previous SSK works the aim of focusing on milieus and networks. Whilst SSK highlighted the falsehoods of the "Great Man of history" approach, it simultaneously demonstrated to biographers of science that a scientific actor, when placed in social, cultural, political, and economic contexts, could be a legitimate object of study and a part of the historiography of science.

Historiographical debate about biography and individuals returned during the 1980s following the defence of biographical approaches put forward by Thomas L. Hankins.[14] Hankins elaborated the primary dilemma, as he saw it, with biographical writing, noting that 'many scientists are not suitable subjects for biography' and that writing biographies of what he deemed "suitable" persons 'distorted the dimension of history by focusing on the head table and ignoring the other banqueters'. He supported the notion that there will always be a place for biographies of celebrated scientists such as Darwin, Einstein, and Faraday. However, beyond this Hankins argued that biography was unsuitable for studying the social and institutional organisation of science, nor was it 'the proper mode for describing the development of a field of science through time'. He concluded that 'a fully integrated biography of a scientist which includes not only his personality, but also his scientific work and the intellectual and social context of his times, is still the best way to get at many of the problems that beset the writing of history of science'. This was an early argument within the historiography of science for what is now termed "sociological biography".[15]

According to Charles Thorpe, an advocate of sociological biography, this approach allows us to see individuals as "exemplars" of their age, providing a key sociological understanding of scientific relations in a given time through the history of individuals. This approach exemplifies social habits through individual characters. Although there is a growing acceptance of sociological biography amongst historians of science, it remains a contested field of enquiry within the discipline. In their sweeping biographical study of Lord Kelvin, Crosbie Smith and Norton Wise leaned heavily on new cultural history approaches then entering the field, rather than attempting to combine sociological studies of science with biographical methodologies.[16] Placing William Thomson (Lord Kelvin) at the centre of their narrative, they used his career as the foundation upon which they carefully constructed a social and cultural history of late nineteenth-century science and technology in Britain. They also broke with biographical tradition in that they did not use the birth and death dates of Kelvin for periodisation, instead only beginning with his education at Cambridge and talking about his early life through the narrative of his father's life. In much the same way, this book concerns itself only with the later career of George Deacon, from his entry into the Royal Navy scientific divisions in 1939 through to his retirement in 1971, and it uses this career to analyse a much broader historical canvas. In any case, this study should not be construed as a social biography, since the goal is not to examine Deacon as an "exemplar" of his time, but rather as someone who established the relevant connections that allowed British oceanography to thrive.

In recent decades, there have been further attempts to provide a theoretical backbone, 'a framework to conceptualise the "singularity" of individuals', and reflect the historiographical return to the problem of individual actors and biography.[17] In the 2006 introduction to a special *Isis* issue on "Fragmented Lives", Joan Richards argued that 'scientific biographers must locate their subjects in a complex landscape'.[18] The collection of articles articulated that the history of science must avoid becoming an impersonal study of institutions because there were significant questions to be answered about 'the nature of leadership in a scientific community and the moral courage of a scientific life'.[19] As part of this reinterpretation of the role of actors within the history of science, Ronald E. Doel articulated the notion of the intersectional actor, between the world of science and state structures. Doel characterises this as 'science in the black'. Doel argued that, especially during the Cold War, large scientific networks were created and administered by individuals who were skilled in more than one area of expertise.[20] These individuals united, for instance, scientific dexterity with a good knowledge of political contingencies and ways to manage large groups. In the context of the Cold War this was especially important as some of these "hybrids" could play key roles in international relations, helping to overcome political divisions by using science as a way to instigate collaborative work. Or they could rise to leading roles by offering, either in the open or in secret, vital pieces of intelligence to government officials.[21] Doel's intersectional actor is not merely an exemplar, as in sociological biographies. It is a scientist that possesses a variety of skills, enabling them to shape the historical trajectory of networks in order to "get their own way" (as Deacon did).

The notion of a multifaceted, "hybrid", historical actor is vital to this book, and is one of the most significant methodological advantages of producing biographical narratives centring on Cold War scientists.[22] A. Hunter Dupree identified the Second World War as the moment when scientists entered the realm of foreign policy in unprecedented numbers.[23] The broadening of scientists' roles and their spheres of influence has also been well documented by Chandra Mukerji, Gary Weir, and Jacob Hamblin in their accounts of Cold War oceanography.[24] However, there are only a few histories of non-American hybrid scientific actors. Most look at individual histories of "significant" scientists whom we may consider "exemplars". One such case is Mary Jo Nye's biography of Patrick Blackett.[25] Whilst there has been an attempt to highlight the role of individuals as "persuaders" in the development of British post-war science policy, placing these actors' interconnections in historical context has proven problematic.[26] This book draws together the understanding of the hybrid scientist at the highest levels of policy making in the British state, and the networks that underpinned their ability to have a leading role.

This hybridity of an actor, the interconnections between their scientific work and the intellectual, political, and cultural fashioning of their identity is exactly, in my view, what sociological biography should strive for. On the other hand this approach is challenged by writers of traditional biographies such as Leslie Pearce Williams, who wrote a series of scientific biographies during the 1960s, and has described Bruno Latour's *The Pasteurization of France* as 'Hamlet without the Prince of Denmark', arguing that 'Latour systematically depasteurizes Pasteur's achievement'.[27] This criticism misses the essential difference between traditional biography and the new use of biography in the history of science: whereas biography had concerned the individual's life and its "lessons", modern biography sees an actor's career trajectory as a device that allows the historian to approach questions about the impact of individual actors on both the organisation of science and intellectual life more generally.[28] This is the use to which I have put George Deacon's career, paying greater attention to the dynamic of bureaucratisation and the closer integration of ocean science into the affairs of the British state than to Deacon's scientific achievements as such. Deacon's hybridity as an actor, his involvement in science, the political administration of science, and the military governance of science, allows this book to study the dynamics of ocean science in the British state through the lens of his networks. This hybridity was embedded in the entanglement of networks that Deacon contributed to establishing and developing between scientists, civil servants, and military officers, with him at the centre acting as the overlap and connection between these separate spheres.

To understand the role of the central actor of this book, George Deacon, in this "sea" of scientific-political-military affairs requires careful consideration of the roles, dynamics, and influence of networks on an actor-to-actor basis and the outcomes for ocean science that resulted.

Networks of Sea Power

A second central argument of this book is that, having been appointed to a position of power and prestige within the community of British oceanographers, George Deacon worked consistently to consolidate and maintain his position as the main negotiator for ocean science within government. Furthermore, members of the civil service and actors within the Admiralty administration assisted Deacon with this hegemonic positioning. Why one individual should come to so dominate scientific-political relations within a particular scientific discipline might have been explained by early biographers of science as the result of Deacon's charisma, personality, or unique skill set.[29] I argue instead that this might have more to do with mutual interdependence. Deacon became the mediator allowing some of his patrons to achieve what they wanted; whilst Deacon "got his own way", this was in essence "their own way".

By analysing an actor's agency, it is possible to discern the dynamics at play within an identified network.[30] The study of networks from this perspective is not novel within the historiography of science that is concerned with the scientific "community", nor in intellectual history, which often returns to the study of republics of letters amongst scholars in various periods.[31] Where this book departs from these studies is in considering *interconnecting* networks, rather than remaining statically within an easily definable group of actors whose common interest drives an internal network dynamic. What is at stake here is an external dynamic between networks, driven by the objective of shaping a particular disciplinary trajectory—in this case oceanography. This is not a novel approach in other academic communities, such as the study of public administration and policy, where the notion of *policy networks* has spawned a similarly complex literature.

Policy is not made in political confrontations in Parliament, or in electoral contests, but in an underworld of committees, civil servants, professions, and interest groups.[32] This is the policy network: a world below the public level of government that has been referred to by some political scientists as the 'sub-government'.[33] Policy networks are conceptualised as a 'horizontal coordinating process in which a stable and lasting relationship is formed between government actors and private actors, who together share a common policy focus'.[34] This is the world in which Deacon operated, networking with oceanographers (scientists), civil servants, and military administrators (who were also known as civil servants in the British system, but delineated here for clarity). Political scientists William Coleman and Grace Skogstad have further clarified this definition: separating the notions of policy community and policy network, they define a policy network as 'the properties that characterize the relationship among the particular set of actors that forms around an issue of importance to the policy community'.[35] They see community as referring to a general set of actors, whereas network refers more specifically to the relationship and interaction between a specified set of actors from within the larger community.

The debate within the field of political science regarding policy networks is often more concerned with developing typologies into ontologies than it is with the actualities of policy network dynamics.[36] Approaches towards studies of network governance have shifted, which in turn has resulted in a reorientation of focus within networks studies to individuals.[37] Networks exist only because of interactions between individual actors; what differentiates policy networks from social network analysis is that in a governmental setting these actors interact because of an interdependence that exists between them, although that interdependence may take many forms.[38] Whilst actors may be said to be at the centre of a network, there remain many dimensions to any network interaction. Political scientist Frans van Waarden defines these dimensions as '(1) actors, (2) function, (3) structure, (4) institutionalization, (5) rules of conduct, (6) power relations and (7) actor strategies'.[39] I argue that a policy network existed during the Cold War between British oceanographers, naval officers, and civil servants, and that we can use biographical elements of key actors in this network to explain the remaining dimensions of van Waarden's model of network interaction.

So why did a policy network exist in Cold War Britain that underpinned the development of ocean science? The main feature of any relationship between the machinery of government and external actors is interdependence. From the perspectives of government administration, this is because, as van Waarden argues, 'administrators need political support, legitimacy, information, coalition partners in the competition with other sections of the bureaucracy, and assistance in the implementation of policy'. Although van Waarden was talking of civil servants, many of these needs are just as applicable to scientists and military officers like Deacon. Also noting that costs are cut through the use of interest groups (the cost of 'repeated effort') and administrators (the 'costs of information collection'). The "cost" here is not economic, rather it is grounded in prestige, legitimacy, and scientific cohesion. Debate within an interest group (in this case scientists) was difficult for government officials to mediate; however, building Deacon into a spokesman for oceanography streamlined the administrative process in favour of government/military positions. Essentially this is the central function of a network: to establish efficient channels of communication between different communities.

Whilst a study of the interdependency of science and government within policy networks has its merits, the channels of communication within policy networks can be useful in illuminating policy processes. Ultimately, policy networks are about actions and 'the importance of the process of bargaining, coalition formation, and conflict'.[40] However, understanding what the network achieves does not necessarily shed light on *how* or *why* the network acts in certain ways. These questions have more to do with structures, so that the "action" of a network is based on two factors: the explicit motive and the implicit effect of the structure connections.[41] Existing strands of literature have identified patronage and resources as the chief motive driving the scientists, an interpretation that this book shows to be correct, but simplistic.[42] For the military officers, oceanographers were a key group of workers improving anti-submarine warfare techniques and contributing to national defence, whilst simultaneously raising awareness of the continuing British connection to the oceans for civil servants in Whitehall. The civil servants' motivations are less clear: having the least to gain from the interaction, civil service interaction is motivated primarily by political policy—the drive to use science as a solution to national problems, not merely defensively but also economically. In essence, the network "acts" because of the multiplicity of motives at play.

Scientists who interact in the policy network are therefore mainly those who stand to benefit the most from such interconnections.[43] This is why senior scientists, like Deacon, who had established scientific careers are mainly discussed in this work. Established scientists' roles centred on securing future funding, maintaining current funding, and establishing links in order to access resources; whilst their juniors may have shaped the field scientifically, their ability to do so was predicated on the ability of superiors to supply facilities and enable this work. Similarly, in the military network it was not the senior officers from the Board of the Admiralty who were engaging in conversations with oceanographers, nor its active scientists, it was the military's own scientific administrative branches who engaged in this dialogue. Yet through their connections to other parts of the military establishment, they in turn communicated the needs of civilian oceanographers up the chain of command. Therefore, by association the tie between civilian science and military science opened up a link from civilian science to the military elites, albeit though another group of intermediaries.

The civil service is the most transient part of the interconnected policy network set out here, although the major departments of state, the Cabinet Office, and the Treasury loom large as the centre of political policy on the one hand, and the holders of the government purse on the other. Civil servants often switched departments or changed jobs frequently and this makes the study of these nodes through time very difficult. The names of civil servants and the parts of government they serve shift almost continuously throughout this book, and are therefore carefully accounted for. Thus we return to the problem of actors, and it is easy to see why, in such a transient organisation, giving agency to the actors rather than to institutions or committees enriches our understanding of the bureaucratic process.

In order to deal with the transient nature of the civil service network over time, and the multi-level nature of the military network, the analytical gaze of this work falls on the relatively stable network of oceanographers in the UK. The personal papers of George Deacon are a useful grounding for this work, because of his continued role as an oceanographic "leader" and hence his influence with the military and the state. As Dominque Pestre has suggested, studying "actors" inevitably results in the selection of a very specific group for study, which produces very specific ontologies. The group of actors which this study *could* have selected to be the object of study is large, therefore this book is situated within a small territorial, disciplinary, and national stage or "scene of inquiry". As Pestre has summarised,

> In a strong sense, we never *follow* actors. Rather we *select* them, we select entities that will play a role on the stage of our story…we choose the ones which will be pertinent to what we care about. Symmetrically, we ignore many others – and, explicitly or not, our choice reveals what is at stake for us and frames the conclusions that can be drawn from our story.[44]

Throughout this book Pestre's comments have been kept in mind. While the use of an innovative methodology means that *Ocean Science and the Cold War State* pioneers the use of a bio-network approach in the historiography of British oceanography, it is equally true that other actors could have been selected and different narratives elaborated. Therefore, *Ocean Science and the British Cold War State* is not so much a history of British oceanography as a history of the policy networks between ocean science and the British state.

Cold War Oceanography: Literature and Interpretations

There is a distinct lack of histories of twentieth-century British ocean science. Those that do exist are primarily based either on personal reflections, published accounts, or scientific papers; archival sources are not generally consulted, and even where they are it is rare for records beyond a single scientist's private papers to be discussed. The two most recent examples of this are Margaret Deacon, Tony Rice, and Colin Summerhayes' 2001 book *Understanding the Oceans: A Century of Marine Exploration,* and Anthony Laughton, William J. Gould, Tom Tucker, and Howard Roe's 2010 book, *Of Seas and Ships and Scientists: The Remarkable History of the UK's National Institute of Oceanography, 1949–1973.*[45] These accounts not only suffer from the imbalance identified above, they are also predominantly hagiographical, written with the express intention of "celebrating" the achievements of British ocean scientists. Nevertheless, they provide a useful framework to understand the scientific undertakings of British ocean scientists, as well as identifying the individuals who collaborated to produce this research.

Set against these collective narratives there are a few academic treatments of the history of modern ocean science in Britain. Helen Rozwadowski provides a history of nineteenth-century British oceanography in *Fathoming the Ocean: the Discovery and Exploration of the Deep Sea,* which is both a scientific and cultural history. Rozwadowski envisions the ocean as a new frontier both figuratively and in the popular imagination; in doing so providing one of the few academic accounts of the origins of the *Challenger* expedition, which is more commonly cited by the aforementioned treatments as the beginning of the age of modern oceanography, born of British pluck. More recently, Anna Carlsson-Hyslop has ventured into the twentieth century with an account of the development of applied physical oceanography at the Liverpool Tidal Institute during the first half of the century.[46] However, these accounts do not focus enough on the political framework in which oceanography thrived due to its military connections, imperial basis, and national/geopolitical security implications. These studies are useful for this work in that they show that certain themes recounted in the promotion of ocean science by oceanographers have long roots, based in these earlier accomplishments of military-scientific and scientific-industrial collaboration respectively.

Whilst political frameworks are conspicuously absent from British-centric accounts, they are well placed at the heart of accounts of oceanography in the United States, reflecting the importance attributed to the earth sciences during the Cold War. There is a sophisticated literature concerning the development of US oceanography as a Cold War science. This historiography is primarily concerned with mapping the experience of American oceanographers within the historical framework of the Cold War.[47] These accounts universally argue that the Cold War shaped, and was shaped by, developments in the oceanographic sciences, domestically and internationally. These histories seek to site oceanographic sciences in four primary contexts: military, political, international relations, and secrecy and science.

The military context of oceanographic research has received significant attention. Historians Ronald Rainger, Gary E. Weir, Naomi Oreskes, Jacob D. Hamblin, and David K. Van Keuren have all discussed the link between oceanography and the US Navy.[48] As this discussion of the links between military and scientific enterprise during the Cold War has developed, a more nuanced appreciation of the shape of the relationship has come to light. Therefore although Rainger sides with Paul Forman's "distortionist hypothesis" in seeing oceanographic sciences as shaped by the military in response to the needs of patronage and the pressures of the Cold War, Weir has suggested that their interaction should be seen more as a partnership.[49] However, in setting out the military-scientific interactions in this way, Weir struggles to pin down the power relationship between the two groups. This challenge was picked up by Oreskes, who analysed what they termed the 'context of motivation', arguing that each side provided a justification for the activities of the other. The relationship, seen in this light, becomes mutually supporting. Going even further, Hamblin suggests that oceanography provided the perfect cover for specific naval research tasks. Highlighting the "innocence" of fundamental research, or basic science, and also its duplicity, suggesting in later work that the same data was useful to both the military and scientists, although the use to which each put this knowledge differed greatly.[50] These studies have illuminated the importance of the military–science relationship of the Cold War, which reveals a remarkable expansion of interactions after the Second World War. Following Deacon's role in the policy and scientific networks helps to understand the dynamics of patronage in British oceanography. This connection between the military uses of oceanography and the close patronage relationship between the two during the early Cold War raises the question of the extent to which the Navy distorted ocean science.[51]

This is a question that has been posed many times before in the case of US science, but is also relevant in the British case, where the predominant model is David Edgerton's British Warfare State.[52] On the one hand, this approach raises the question of whether the military distorted science from its "true" trajectory or was merely supplementing, albeit generously, pre-existing trajectories.[53] As has been noted by many scholars and recently in an edited volume on geoscience, surveillance, and the Cold War, scientists had a great deal of agency within military-scientific relationships.[54] Whilst military patronage certainly shaped the environments in which research questions were chosen and directly funded most of the research undertaken, one should never lose sight of the degree to which *both* the military and the scientists aimed to maximise their investment, be it of money or expertise, in order to further their own ambitions, which were sometimes complementary but often divergent.

Sociologist Kelly Moore has analysed how scientists lobbied both for and against the use of science for military activities as "political actors".[55] Through a critical analysis of the ties between science and the military at the level of individuals, Moore demonstrates the extent to which scientists operated as free agents and yet were "bound" to the state. By problematising the idea of 'bargaining with the devil', as Moore caricatures relations between 1945 and 1970, revealing the duality of the relationship and the seeming lack of moral or ethical issue this engendered in the majority. Moore's wider thesis is that although only the minority voiced opinions challenging the status quo, their voices were potent and over time they did not so much bring the whole system down as force change on a small scale, as scientists increasingly joined environmental, anti-war, anti-nuclear movements during the 1960s. Here it is not Moore's arguments I wish to take up so much as the methodology of studying the politics of science–military ties during the period. Throughout this book these formal (and perhaps more importantly, informal) ties are shown to reveal a lot about the course of science in the Cold War. Focusing on bio-networks has revealed that scientists were not only complicit in the use of their science for military purposes, but actively promoted the military applications of their work to political bodies both nationally and internationally to attract funding and resources.

Breaking away from a national approach was the primary goal of Jacob Hamblin's 2005 book, *Oceanography and the Cold War*.[56] In the introduction to this monograph, Hamblin noted the disparity between considering the oceans and their study as a global endeavour and the national histories that dominated the historiography of its more militaristic pursuits. Hamblin demonstrated that, through international committees, American oceanographers attempted to recruit international colleagues to their "big" collaborative projects by distributing Navy funds. In carrying out a shared global data project, many oceanographic projects worldwide could be unwittingly mobilised into fighting the US Navy's Cold War. Hamblin's project situated the international within the context of the US experience because of the primary archives upon which the study was based, but also the preponderance of US oceanographers in international committees during the period. This was in stark contrast to the other main study of twentieth-century international oceanography, Helen Rozwadowski's 2004 book on the International Council for the Exploration of the Seas (ICES).[57] ICES is interesting object of study because it undertook oceanographic research without US involvement: the United States having never joined the organisation. Rozwadowski's study of an intergovernmental panel for the marine sciences by its very nature lacked a firm link back to national developments. This work also took as its central empirical basis the study of the activities of ICES (a body set up to study fisheries), which although innovative in the period before the Second World War, was detached from the Americanisation of international oceanography post 1945.[58] These studies both prompt the serious question of the national, international, and transnational nature of Cold War oceanography, but we have to look beyond this historiography to appreciate the advantages, disadvantages, and challenges of writing the history of a global science discipline in the Cold War period. Here a bio-network approach identifies Deacon as using international networks to boost his role as a major figure in national networks. As the NIO Director he essentially used the influence he gained in international collaborative projects as a lever to influence policy decisions at home, thus shaping the evolution of the oceanography policy network to which he was a contributor. In adding international to existing national links he aimed, essentially, to reinforce his role as the central node of his network.

Recent studies of international science have suggested that rather than seeing nations as providing the foundation of international science, we should also see that international science is constructed in a way that influences national developments. Flows of goods, people, ideas, words, capital, power, and institutions are multidirectional and not merely a two-way process in and out of national contexts.[59] It has been argued that historians of science often either take national and transnational perspectives for granted, or choose to keep them out of sight and in the background.[60] However, a history of ocean science examines a discipline which itself focuses upon spaces nominally "devoid" of national boundaries, where fences and barbed wire cannot enclose territory, and the area is vast. Furthermore, if "science" transcends *national* borders, it is quintessentially *transnational* in character. Though the concept of a *nation* has been well established since the early modern period, by contrast the *international* is used as a reference point directed towards actions, institutions, or even ideologies which have agency at a level above the nation state. In the Cold War perhaps the ultimate international body was the United Nations, but the international cannot exist without the tacit support of the nation state. The *transnational* on the other hand is the study of connections, encounters, and circulations of all types; it is methodologically concerned with the study of multiple forms of interaction, through the consideration of various 'actors'.[61] This need not be limited to individuals but, as Eric Vanhaute has argued, should also include 'institutions, societies and human systems'.[62] Therefore if we take the premise of Mark Walker that 'all science is either national or transnational, and most is both', one may conclude, as Lewis Pyenson does, that boundary study on the basis of nationality is essentially futile.[63]

Yet there is a paradox with which this book grapples: whilst recognition of scientific credibility was increasingly determined in the international community of science during the Cold War, the funding for that very same research was reliant on scientific funding and institutions that were firmly rooted in national contexts.[64] This was clear with the International Geophysical Year, a project that despite its international focus was fundamentally reliant on national financing. Using international prestige to build projects allowed Deacon to lever financial backing from national patrons. On this basis, certain aspects of this study are transnational in character, even if superficially it appears to be the study of a specifically national narrative. Its transnationality comes from the international arenas in which its central actor played a part, but also the way in which oceanography grew as an interconnected discipline, based around collaboration and interchange of resources and scientists during the Cold War. This is achieved through the study of Deacon as at the centre of a bio-network of actors who operate in various contexts, sometimes national, other times international, and more often than not working on both levels simultaneously. In this way it always returns to the paradox of national commitments in transnational systems of scientific governance, and in this way it is a transnational history of a national challenge: maintaining international prestige in a competitive international scientific discipline. There are, however, consequences of situating the narrative in this transnational setting.

One repercussion of the international dimension of oceanography was the issue of scientific secrecy that emerged from the connection between science and national security. This issue first arose in the United States during the Second World War, when the leaders of US oceanography (who were primarily Scandinavian) applied for security clearance to participate in Navy contracts that were being awarded to their institutions.[65] The post-war atomic tests at Bikini Atoll in the Pacific brought oceanographers into the world of nuclear security and although many scientists were by this time out of uniform, the tests not only brought civilian science into collaboration with military testing, but also gathered oceanographers inside the "black world" of secrecy.[66] For many oceanographers, the veil of secrecy mattered only if they were the leaders of institutions or directly involved in collaborative projects with the military; this work was under the banner of the "need-to-know" approach. This naïve view was challenged in the case of the Heezen-Thorpe map, in which vast data was collected on the topography of the North Atlantic seabed. Scientists were free to discuss what they were doing, but instead of publishing their results, the Navy (who had funded the expedition) would only allow the release of a map that was devoid of many of the data points collected. Oceanographers from outside the project were forced to either assume that the project had been compromised because of military secrecy or that the science had been nowhere near as accurate or thorough as claimed.

This book reveals that in these situations Deacon acted as the interface between secret and non-secret networks in British and international policy, as was most clearly seen in Deacon's international work for NATO (non-secret) and domestic work for Admiralty research establishments which was secret. Control here manifested itself in regimes of monitoring and surveillance of enemy (and potential enemy) ship movements, and fitted into a larger scheme of monitoring the earth that encompassed the geosciences more generally.[67] This book connects this surveillance dimension to existing approaches that note the connection of ocean science with the military, government, international regimes of scientific governance, and histories of the dark world of secret science, through its study of bio-networks.

The Cold War Surveillance Imperative, Security Dilemma, and Intelligence Challenge

The Cold War was a state of political, military, and cultural tension between nations of the Western alliance—the United States, NATO allies, and others who aligned themselves with these nations—and powers in the Eastern bloc—the Soviet Union and its allies. It emerged after the Second World War and shaped global relations until the period between 1989 and 1991 when the Soviet system collapsed.[68] To put the Cold War at the centre of any post-Second World War history is to regard geopolitics as central to recent history.[69] Essentially, understanding how science became intertwined with geopolitics during the twentieth century is fundamental to appreciating the place of ocean science in the wider Cold War.

Traditionally the Cold War has been primarily framed as a binaristic conflict. This two-dimensional orientation is itself supported by propaganda-led interpretations of East and West in terms of light and dark, right and wrong, free and shackled, just as the Second World War had differentiated between Allied and Axis powers.[70] Post-Cold War-era scholarship has largely abandoned these simplistic narratives, instead building a framework that sees the conflict as global.[71] This is constructed through the agency of newly independent nations free of imperial rule that are seen as indicative of the hegemonic nature of the United States as a new non-imperial actor yet operating in such a way as to influence national trajectories, especially in Western Europe.[72] One way in which the United States attempted to influence national strategic trajectories away from their own imperialistic endeavours towards building what the United States envisaged as a closed world was through the enrolment of Western European states in its surveillance, monitoring, and intelligence-gathering project. This approach to foreign policy by the USA has often been framed through the lens of a security dilemma, but rarely has this been linked to surveillance activities.[73]

The Cold War was predominantly concerned with the remote control of spaces; replacing boots on the ground with infrastructures of surveillance. It also expanded the range of spaces that nations (or superpowers) could control. No longer was national security maintained through geostrategic control of land areas; nations in general, and superpowers specifically, needed to monitor the whole earth from the seabed to outer space. Therefore in conceiving security as a geographic problem as well as a military one, gathering information on enemies became intimately linked with gathering information about the earth.[74]

In this way, the theme of surveillance ties science, politics, and the military together during the Cold War.[75] A new value was placed upon knowledge of the earth itself by states that were increasingly embarking on a quest to know more about those who lived on earth by asking questions of the physical places they inhabited. It has been argued that 'surveillance networks owe their existence, or at least their sophistication and extent, to the dramatic expansion of funding to the geosciences after 1945'; oceanography was one of these sciences that dramatically benefited from such a geopolitical outlook.[76] The link between oceanography and the military was not formed in the Cold War; as Gary Weir has shown, it began much earlier, but because of Cold War data sharing the links between oceanography, surveillance infrastructures, and the military became much more embedded.[77] This gave data a dual role: knowledge of the oceans was a key building block of national security, which in turn provided justification for the expansion of marine science. The desire for global coverage, universal knowledge, and complete surveillance meant that the partial data sets that one nation alone could produce had little value. Rather than restrict the acquisition of global knowledge of the oceans by developing such sensitive information only in what has been called "science in black" (the world of classified science), the USA drove the international sharing of data, even if this meant providing sensitive data to rivals, because a complete global data set was perceived to be almost priceless.[78]

Hybrid actors such as Deacon were central middlemen in the widespread spying *on* science, which provided intelligence on foreign scientific capabilities. Western governments urged their oceanographers to "judge" Soviet science to ensure that scientific claims were realistic and plausible and scientific capabilities correctly appraised.[79] In the United States this led the US IGY committee to send all information it received to the CIA Office of Scientific Intelligence, whereas in Britain the information was collected by the Joint Intelligence Bureau (JIB), who reported to the central clearing house for the British intelligence community, the Joint Intelligence Committee (JIC).[80] The treatment of Deacon here fits this figure, because of his connections and ability to relate to civil servants, military officers, and intelligence specialists and their needs. This group of actors came to shape the development of ocean science not from the laboratory bench but from their office desks as they defined the patronage relationship between the scientists and the state.

There have been calls to assimilate the history of these political administrators of science with the contemporaneous history of the scientific developments to which they minister. Bruno Latour concludes his seminal work *Science in Action* by stating that to understand the work of scientists, one must also understand the work of the "pen-pushers" who organise resources for science, as this is fundamental to any account of modern science.[81] This book discusses the work of George Deacon as he organised resources for British oceanography: as the director of his institute; in Whitehall; in the Admiralty developing anti-submarine warfare techniques; whilst recruiting other scientists to this line of work; or even through top-secret security clearance work. These histories have to be written together, placed in the context of all their separate but interconnected networks, and seen as mutually constructed co-narratives that allow Deacon to be Latour's "pen-pusher" or a hybrid actor; the one who "got things his own way". In this way we get the curious history of how a Cold War career played a part in the shaping of the relationship between ocean science and the British Cold War state.

Using an archive-based approach, this book demonstrates the continued importance of the military in British oceanography from 1945 through to the 1960s. The tangle of bureaucratic politics that science had to weave through to gain a voice in Whitehall, and the role of ocean science and its scientists in the development of surveillance policies and intelligence assessments of Soviet science, are also revealed through a study of the networks in which George Deacon moved. Through the study of networks, the history of British oceanography is placed in an international context, and as this book demonstrates this context can only be analysed in conjunction with a study of national challenges, strategies, and approaches. This book has lessons for the history of science, but also political history, the history of the Cold War, geopolitical history, international relations history, and bureaucratic history. Each of the chapters that follow consequently build this narrative by mapping out a picture of the networks in which George Deacon "sailed", and the impact this had on the development of oceanography in the British Cold War state.

Notes

1. The papers presented at this symposium were reprinted in the *Proceedings of the Royal Society of London: A*, 342:1631 (15 April 1975): 439–591.
2. Edward Bullard, "The Effect of World War II on the Development of Knowledge in the Physical Sciences," *Proceedings of the Royal Society of London. A.*, 342:1631, (April 1975): 531.
3. Ibid., comments on this paper included in the Proceedings.
4. For examples of where this is not achieved see Paul Forman, "Behind Quantum Electronics: National Security as Basis for Physical Research in the United States, 1940–1960," *Historical Studies in the Physical and Biological Sciences*, 18:1 (1987): 149–229; Stuart W. Leslie, *The Cold War and American Science: The Military-Industrial-Academic Complex at MIT and Stanford* (New York: Columbia University Press: 1994); David Kevles, "Cold War and Hot Physics: Science, Security, and the American State, 1945–56," *Historical Studies in the Physical and Biological Sciences,* 20:2 (1990): 239–264.
5. Throughout this book Big Science is invoked in the sense of Big Science as an Instrument rather than Big Science as Politics. For a review of the many uses of "Big Science", see James H. Capshew and Karen A. Rader, "Big Science: Price to the Present," *Osiris* 7 (1992): 8 –9, 12–15. In terms of Big Science as an Instrument, I follow Peter Galison's point that even existing technology (in this case research ships) were exploited so as to realign scientific practice and priorities according to the engineered ethos of Big Science, especially its scientific administrators and committees. Peter Galison, "Bubble Chambers and the Experimental Workplace," in *Observation, Experiment, and Hypothesis in Modern Physical Science,* ed. Peter Achinstein and Owen Hannaway (Cambridge, MA: MIT Press, 1985): 309–373. Peter Galison, "The Many Faces of Big Science," in *Big Science: The Growth of Large-scale Research,* ed. Peter Galison (Stanford, CA: Stanford University Press, 1992): 1–20. In terms of Big Science as Politics I have developed a conceptual framework sited at the level of the policy network which is outlined in section 1.2 of this introduction, rather than seeing Big Science as a rhetorical policy device in its own right.
6. The following definition is adapted from the National Oceanic and Atmospheric Administration definition of oceanography as it appears on their website, www.noaa.gov. Oceanography is a broad discipline made up of various topics, including studies of marine life, zoology and ecosystems, ocean circulation, plate tectonics and the geology of the seabed, and the chemical and physical properties of the ocean. Most oceanographic institutions, however, study these topics in a composite fashion, bringing together the skills of multiple expertise across these approaches. Four categories of

scientists study these aspects, which are often grouped under four distinct sub-groups. Marine biologists, fisheries scientists, and biological oceanographers study plants and animals in the marine environment. Chemical oceanographers and marine chemists study the composition of seawater and its interaction with the atmosphere and the sea floor. Geological oceanographers, marine geologists, and geophysicists concerned with the ocean floor study sea-floor spreading, plate tectonics, and oceanic circulation; they are also interested in the volcanic processes on the seabed. Physical oceanographers, in whom this study is primarily interested, study the physical processes of the ocean, such as waves, currents, tides, and the interactions between the atmosphere and the ocean; they also examine deep currents, the transmission of light and sound through water, and the interactions between one ocean water mass and another. All of these fields can be said to be intertwined and to an extent it is often argued that oceanographers have to have an understanding of all these four branches to study the ocean. This is an abbreviated version of the National Oceanic and Atmospheric Administration website definition of oceanography, http://oceanservice.noaa.gov/facts/oceanographer.html.

7. Thomas P. Hughes used the notion of separate but conjoined spheres of influence in his study of electricity in America. See Thomas P. Hughes, "The Electrification of America: The System Builders," *Technology and Culture*, 20:1 (1979): 124–161.
8. The traditional critique of this was provided by Herbert Butterfield in his seminar on the Whig interpretation of history; Herbert Butterfield, *The Whig Interpretation of History* (London: G.Bell, 1931).
9. David Aubin, Charlotte Bigg, "Neither Genius nor Context Incarnate: Norman Lockyer, Jules Janssen and the Astophysical Self," in *The History and Poetics of Scientific Biography*, ed. Thomas Söderqvist (Farnham, UK: Ashgate, 2007): 54.
10. Robert K. Merton, "Priorities in Scientific Discovery: A Chapter in the Sociology of Science," *American Sociological Review*, 22:6 (1957): 635–659.
11. Trevor J. Pinch, "Sociology of Science," in *Reader's Guide to the History of Science*, ed. Arne Hessenbruch (London: Fitzroy Dearborn Publishers, 2000): 695. Central works in SSK include: Bloor, D., *Knowledge and Social Imagery*, (Routledge: London, 1976); Harry M. Collins, *Changing Order: Replications and Induction in Scientific Practice* (California: Sage, 1985); Bruno Latour, Stephen Woolgar, *Laboratory Life: The Construction of Scientific Facts* (Princeton, NJ: Princeton University Press, 1986); Bruno Latour, *Science in Action: How to Follow Scientists and Engineers through Society* (Cambridge, MA: Harvard University Press, 1987).

12. Michel Callon, "The Sociology of an Actor-Network: The Case of the Electric Vehicle," in *Mapping the Dynamics of Science and Technology: Sociology in the Real World*, ed. Michel Callon, John Law, Arne Rip (London: Macmillan, 1986): 19; Harry M. Collins, Trevor J. Pinch, *Frames of Meaning: The Social Construction of Extraordinary Science* (London: Routledge: 1982); Steven Shapin, "The Politics of Observation: Cerebral Anatomy and Social Interests in the Edinburgh Phrenology Disputes," in *On the Margins of Science: The Social Construction of Rejected Knowledge,* ed. Roy Wallis (Keele, UK: Keele University Press, 1979): 139–178; Donald MacKenzie, "Statistical Theory and Social Interests: A Case Study," *Social Studies of Science*, 8:1 (1978): 35–83.
13. Steven Shapin, Simon Schaffer, *Leviathan and the Air Pump: Hobbes, Boyle, and the Experimental Life* (Princeton, NJ: Princeton University Press, 1985); Andrew Pickering, *Constructing Quarks: A Sociological History of Particle Physics* (Edinburgh: Edinburgh University Press, 1984).
14. Thomas L. Hankins, "In Defense of Biography: The Use of Biography in the History of Science," *History of Science*, 17 (1979): 1–16.
15. Charles Thorpe, *Oppenheimer: The Tragic Intellect* (Chicago: Chicago University Press, 2006): 11, 13–14; and Simone Turchetti, *The Pontecorvo Affair: A Cold War Defection and Nuclear Physics* (Chicago: Chicago University Press, 2012): 9 (and throughout the work).
16. Crosbie Smith, M. Norton Wise, *Energy & Empire: A biographical Study of Lord Kelvin* (Cambridge: Cambridge University Press, 1989).
17. Aubin, Bigg, "Neither Genius nor Context Incarnate," 56.
18. Joan L. Richards, "Introduction: Fragmented Lives," *Isis*, 97:2 (2006): 302–305.
19. Theodore M. Porter, "Is the life of the Scientists a Scientific Unit?," *Isis*, 97:2 (2006): 314; Mary J. Nye, "Scientific Biography: history of Science by Another Means," *Isis*, 97:2 (2006): 325; Mary Terrall, "Biography as Cultural History of Science," *Isis*, 97:2 (2006).
20. Ron E. Doel, "Scientists as Policymakers, Advisors, and Intelligence Agents: Linking Contemporary Diplomatic History with the History of Contemporary Science," in *The Historiography of Contemporary Science and Technology,* ed. Thomas Söderqvist (Reading, UK: Harwood Academic Publishers, 1997).
21. Doel, "Scientists as Policymakers," 215.
22. Here I use hybrids, because I do not want to confuse the use of intersectional here with the notion of intersectionality now in use by social scientists writing since Doel's earlier use of the term. See Patricia Hill Collins, "Intersectionality's Definitional Dilemmas," *Annual Review of Sociology,* 41 (2015): 1–20.
23. Doel, "Scientists as Policymakers," 215.

24. Chandra Mukerji, *A Fragile Power: Scientists and the State* (Princeton, NJ: Princeton University Press, 1989); Gary E. Weir, *An Ocean in Common: American Naval Officers, Scientists, and the Ocean Environment* (College Station: Texas A&M University, 2001); Jacob D. Hamblin, *Oceanographers and the Cold War: Disciples of Marine Science* (Seattle: University of Washington, 2005).
25. Mary J. Nye, *Blackett: Physics, War, and Politics in the 20th Century* (Cambridge, MA: Harvard University Press, 2004) and Mary J. Nye, *Michael Polanyi and His Generation: Origins of the Social Construction of Science* (Chicago: University of Chicago Press, 2011); for a non-British example see Peder Roberts, "Intelligence and Internationalism: The Cold War Career of Anton Bruun," *Centaurus*, 55:3 (2013): 243–263.
26. Alexandros Oikonomou, "The Hidden Persuaders: Government Scientists and Defence in Post-war Britain," (PhD diss., Imperial College London, 2011).
27. Leslie Pearce Williams, "The Life of Science and Scientific Lives," *PHYSIS*, (1991): 199–213.
28. Thorpe, *Oppenheimer*, 4.
29. Recent historiography has problematised this notion; see Charles Thorpe, Steven Shapin, "Who was J. Robert Oppenheimer? Charisma and Complex Organization," *Social Studies of Science*, 30:4 (2000): 545–590; see also Chapter 4, "Charisma as a 'Career'," in Andrew D. McCulloch, *Charisma and Patronage: Reasoning with Max Weber*, (Farnham, UK: Ashgate, 2014).
30. Jenny M. Lewis, "The future of network governance research: strength in diversity and synthesis," *Public Administration*, 89:4 (2011): 1224.
31. Jack Morrell and Arnold Thackray, *Gentlemen of Science: Early Years of the British Association for the Advancement of Science* (Oxford: Oxford University Press, 1981); Dena Goodman, *The Republic of Letters: A Cultural History of the French Enlightenment* (Ithaca, NY: Cornell University Press, 1994).
32. Marsh, Rhodes, *Policy Networks*.
33. Grant Jordan, "Sub-Governments, Policy Communities, and Networks: Refilling the Old Boots," *Journal of Theoretical Politics*, 2:3 (1990): 319–338.
34. Gila Menahem, "Policy Paradigms, Policy Networks and Water Policy in Israel," *Journal of Public Policy*, 18:3 (1998): 284.
35. William D. Coleman and Grace Skogstad, "Policy Communities and Policy Networks: A Structural Approach," in *Policy Communities and Public Policy in Canada*, ed. William D. Coleman and Grace Skogstad (Toronto: Copp Clark Pitman, 1990).

36. Frans van Waarden, "Dimensions and types of policy networks," *European Journal of Political Research*, 21:1 (1992): 29–52; Erik hans Klijn, "Analysing and Managing Policy Processes in Complex Networks: A Theoretical Examination of the Concept Policy Network and its Problems," *Administration & Society*, 28:1 (1996): 90–119; Erik hans Klijn, Joop Koppenjan, "Governance Network Theory: Past, Present, Future," *Policy & Politics*, 40:4 (2012): 587–606.
37. Jenny M. Lewis, "The Future of Network Governance Research: Strength in Diversity and Synthesis," *Public Administration*, 89:4 (2011): 1224; Klijn, Koppenjan, "Governance Network Theory," 594.
38. Waarden, "Dimensions and Types of Policy Networks," 31; Klijn, "Policy Processes in Complex Networks," 94; Klijn, Koppenjan, "Governance Network Theory," 591.
39. Waarden, "Dimensions and Types of Policy Networks," 32.
40. Klijn, "Policy Processes in Complex Networks," 94.
41. Heather Creech and Terri Willard, "Strategic Intentions: Managing Knowledge Networks for Sustainable Development. How to Manage a Successful Knowledge Network" (Winnipeg, Canada: International Institute for Sustainable Development, 2001): 82–88. http://www.eldis.org/go/home&id=30144&type=Document#.VKC3tAA8.
42. As shown by Dominique Pestre, historians of science have rarely studied the interactions of scientists and politicians, preferring to speak of the patronage of government rather than the deals with the state. Dominique Pestre, "Science, Political Power and the State," in *Companion to Science in the Twentieth Century,* ed. John Krige and Dominique Pestre (London: Routledge, 2003): 62–63, 65.
43. Pestre, "Science, Political Power," 70.
44. Dominique Pestre, "Debates in Transnational and Science Studies," *British Journal for the History of Science* (2012): 435.
45. Anthony Laughton, John Gould, M.J. 'Tom' Tucker, Howard Roe, *Of Seas and Ships and Scientists: The Remarkable Story of the UK's National Institute of Oceanography* (Cambridge: Lutterworth Press, 2010).
46. Anna Carlsson-Hyslop, "An anatomy of storm surge science at Liverpool Tidal Institute 1919–1959: forecasting, practices of calculation and patronage," (PhD diss., University of Manchester, 2010).
47. Jacob D. Hamblin, Gary E. Weir, Helen Rozwadowski, Gary Kroll, Naomi Oreskes, Roland Rangier.
48. Roland Rainger, "Science at the Crossroads: The Navy, Bikini Atoll, and American Oceanography in the 1940s," *Historical Studies in the Physical and Biological Studies*, 30:2 (2000): 349–371; Roland Rainger, "Patronage and Science: Roger Revelle, The U.S. Navy and Oceanography at the Scripps Institution," *Earth Sciences History*, 19:1 (2000): 58–89; Gary

E. Weir, *An Ocean in Common: American Naval Officers, Scientists, and the Ocean Environment* (College Station: Texas A&M University, 2001); Naomi Oreskes, "A Context of Motivation: US Navy Oceanographic Research and the Discovery of Sea-Floor Hydrothermal Vents," *Social Studies of Science*, 33:5 (2003): 697–742; Jacob D. Hamblin, "The Navy's 'Sophisticated' Pursuit of Science: Undersea Warfare, the Limits of Internationalism, and the Utility of Basic Research, 1945–1956," *Isis*, 93:1 (2002): 1–27; David Van Keuren, "Cold War Science in Black and White: US Intelligence Gathering and its Scientific Cover at the Naval Research Laboratory, 1948–62," *Social Studies of Science*, 31:2 (2001): 207–229.

49. Paul Forman's 'distortionist critique', the notion that the US military (negatively) shaped scientific research agendas, was first called this by Roger Geiger; see Roger Geiger, "Review of *The Cold War and American Science: The Military-Industrial-Academic Complex at MIT and Stanford*," *Technology and Culture*, 34:3 (1994): 629–631; Weir, *An Ocean in Common*.
50. Hamblin, *Oceanographers and the Cold War*.
51. David Hounshell, "The Cold War, RAND, and the Generation of Knowledge, 1946–62," *Historical Studies in the Physical and Biological Sciences*, 27 (1997): 239.
52. David Edgeton, *Warfare State: Britain, 1920–1970* (Cambridge: Cambridge University Press, 2005).
53. Advocating the former are Paul Forman, "Behind Quantum Electronics: National Security as Basis for Physical Research in the United States, 1940–1960," *Historical Studies in the Physical and Biological Sciences*, 18:1 (1987): 149–229; Leslie, *The Cold War and American Science*. A key advocate of the second position is David J. Kevles, "Cold War and Hot Physics: Science, Security, and the American State," *Historical Studies in the Physical Sciences*, 20 (1990): 239–264.
54. See Simone Turchetti, Peder Roberts, *The Surveillance Imperative: the Geo-Sciences and the Cold War* (London: Palgrave, 2014): 9; and the work of Kai-Henrik Barth, Ronald E. Doel, Naomi Oreskes.
55. Kelly Moore, *Disrupting Science: Social Movements, American Scientists, and the Politics of the Military, 1945–1975* (Princeton, NJ: Princeton University Press, 2008).
56. Jacob D. Hamblin, *Oceanographers and the Cold War: Disciples of Marine Science* (Seattle: University of Washington, 2005).
57. Helen Rozwadowski, *The Sea Knows No Boundaries: A Century of Marine Science under ICES* (Seattle: University of Washington Press, 2004).
58. John Krige, *American Hegemony and the Postwar Reconstruction of Science in Europe* (Cambridge, MA: MIT Press, 2006) provides an account of US interest in European scientific development after 1945.

59. Here I have paraphrased the list of movements and forces suggested as possible objects of transnational study by Pierre-Yves Saunier, "Entry: Transnational," in *The Palgrave Dictionary of Transnational History*, ed. Akira Iriye and Pierre-Yves Suanier (New York: Palgrave, 2009): 1047–1055.
60. Mark Walker, "The 'National' in International and Transnational Science," *British Journal for the History of Science*, 45:3 (2012): 363.
61. Dominique Pestre, "Debates in Transnational and Science Studies: A Defence and Illustration of the Virtues of Intellectual Tolerance," British Journal for the History of Science 45, no. 3 (2012), 433.
62. Eric Vanhaute, E., "Who is Afraid of Global History? Ambitions, Pitfalls and Limits of Learning Global History", *Österreichische Zeitschrift für Geschichtswissenschaften*, 20:2 (2009): 27.
63. Walker, "The 'National' in International and Transnational Science,"; Lewis Pyenson, "An End to National Science: The Meaning and the Extension of Local Knowledge," *History of Science* (2002): 251–290.
64. Walker, "The 'National' in International and Transnational Science," 359.
65. Naomi Oreskes and Ronald Rainger, "Science and Security before the Atomic Bomb: The Loyalty Case of Harald U. Sverdrup," *Studies in History and Philosophy of Science Part B: Studies in History and Philosophy of Modern Physics*, 31:3 (2000): 309–369.
66. Ron Doel, "Scientists as Policymakers", 215–222.
67. See the papers in Turchetti, Roberts, "Introduction," *The Surveillance Imperative*.
68. John Lewis Gaddis, *Strategies of Containment: A Critical Appraisal of Postwar American National Security Policy* (Oxford: Oxford University Press, 1982); John Lewis Gaddis, *The Cold War* (London: Allen Lane, 2005); Odd Arne Westad, *The Global Cold War: Third World Interventions and the Making of Our Times* (Cambridge: Cambridge University Press, 2007); Melvyn P. Leffler and David S. Painter, *Origins of the Cold War: An International History*, (Routledge: London, 2005); Heonik Kwon, *The Other Cold War* (New York: Columbia University Press, 2010); Odd Arne Westad, "The Cold War and the International History of the Twentieth Century," in *The Cambridge History of the Cold War*, Melvyn P. Leffler, Odd Arne Westad (Cambridge: Cambridge University Press, 2012); Jonathan Haslam, *Russia's Cold War: From the October Revolution to the Fall of the Wall* (New Haven, CT: Yale University Press, 2012).
69. Akira Iriye, "Historicizing the Cold War," in *The Oxford Handbook of the Cold War*, Richard H. Immerman and Petra Goedde (Oxford: Oxford University Press, 2013): 15–31; also Geoffrey Warner, "The Geopolitics and the Cold War," in *The Oxford Handbook of the Cold War*, Richard H. Immerman and Petra Goedde (Oxford: Oxford University Press, 2013): 15–31.

70. This simplistic construction was particularly prevalent during the 1980s, especially in the work of John Lewis Gaddis, David Engerman, "Ideology and the Origins of the Cold War, 1917–1962," in Melyvn P. Leffler, Odd Arne Westad, *Cambridge History of the Cold War* (Cambridge: Cambridge University Press, 2010); Naoko Shibusawa, "Ideology, Culture, and the Cold War," in *The Oxford Handbook of the Cold War*, Richard H. Immerman and Petra Goedde (Oxford: Oxford University Press, 2013).
71. Odd Arne Westad, *The Global Cold War: Third World Interventions and the Making of our Times,* (Cambridge: Cambridge University Press, 2006); Heonik Kwon, *The Other Cold War,* (New York: Columbia University Press, 2010); Prasenjit Duara, "The Cold War as a Historical Period: An Interpretive Essay," *The Journal of Global History,* 6:3 (2011): 457–480.
72. Prasenjit Duara, "The Cold War as a Historical Period: An Interpretive Essay," *Journal of Global History,* 6:3 (2011): 457–480; Jeffrey G. Giauque, *Grand Designs and Visions of Unity: The Atlantic Powers and the Reorganization of Western Europe, 1955–1963* (Chapel Hill: University of North Carolina Press, 2002); Wilfried Loth, *Europe, Cold War and Coexistence* (London: Frank Cass, 2004); Geir Lundestad, *The United States and Western Europe since 1945: From "Empire" by Invitation to Transatlantic Drift* (Oxford: Oxford University Press, 2003); Marc Trachtenberg, *A Constructed Peace: The making of the European Settlement, 1945–1963* (Princeton, NJ: Princeton University Press, 1999); for science specifically, John Krige, *American Hegemony and the Postwar Reconstruction of Science in Europe,* (Cambridge, MA: MIT Press, 2006).
73. Roger Jervis, "Was the Cold War a Security Dilemma," *Journal of Cold War Studies,* 3 (2001): 36–60; Charles L. Glaser, "The Security Dilemma Revisited," *World Politics,* 50 (1997): 171–201; Michael S. Goodman, *Spying on the Nuclear Bear: Anglo-America Intelligence and the Soviet Bomb* (Redwood City, CA: Stanford University Press, 2007); Charles A. Ziegler, David Jacobson, *Spying without Spies: Origins of America's Nuclear Surveillance System* (Conn.: Greenwood Publishing, 1995).
74. Turchetti, Roberts, "Introduction", *The Surveillance Imperative,* 2.
75. Kristie Macrajis, "Techniphilic Hubris and Espionage Styles During the Cold War," *Isis,* 101:2 (2010): 378–385.
76. Turchetti, Roberts, "Introduction", *The Surveillance Imperative,* 2.
77. Weir, *An Ocean in Common*; also Oreskes, "A Context of Motivation," 697–742.
78. Hamblin, J.D., *Oceanographers and the Cold War* (2005); Elena Aronova, Karen S. Baker, Naomi Oreskes, "Big Science and Big Data in Biology: From the International Geophysical Year through the International Biological Program to the Long Term Ecological Research Program, 1957-present," *Historical Studies in the Natural Sciences,* 40:2 (2012):

183–224; David Van Keuren, "Cold War Science in Black and White: US Intelligence Gathering and Its Scientific Cover at the Naval Research Laboratory, 1948–1962," *Social Studies of Science*, 31 (2001): 207–229.
79. Turchetti, Roberts, "Introduction", *The Surveillance Imperative*, 7.
80. Michael S. Goodman, *The Official History of the Joint Intelligence Committee, Volume 1: From the Approach of the Second World War to the Suez Crisis* (London: Routledge, 2014); Huw Dylan, *Defence Intelligence and the Cold War: Britain's Joint Intelligence Bureau 1945–1964* (Oxford: Oxford University Press, 2014).
81. Bruno Latour, *Science in Action: How to follow Scientists and Engineers through History* (Cambridge, MA: Harvard University Press, 1987): 257.

References

Aronova, Elena, Karen S. Baker, Naomi Oreskes. "Big Science and Big Data in Biology: From the International Geophysical Year Through the International Biological Program to the Long Term Ecological Research Program, 1957-Present." *Historical Studies in the Natural Sciences* 40, no. 2 (2012): 183–224.

Aubin, David, Charlotte Bigg. "Neither Genius nor Context Incarnate: Norman Lockyer, Jules Janssen and the Astrophysical Self." In *The History and Poetics of Scientific Biography*. Edited by Thomas Söderqvist. Farnham, UK: Ashgate, 2007.

Bloor, David. *Knowledge and Social Imagery*. London: Routledge, 1976.

Bullard, Edward. "The Effect of World War II on the Development of Knowledge in the Physical Sciences." *Proceedings of the Royal Society of London. A* 342, no. 1631 (1975): 519–536.

Butterfield, Herbert. *The Whig Interpretation of History*. London: G.Bell, 1931.

Callon, Michel. "The Sociology of an Actor-Network: The Case of the Electric Vehicle." In *Mapping the Dynamics of Science and Technology: Sociology in the Real World*. Edited by Michel Callon, John Law, Arne Rip. London: Macmillan, 1986.

Capshew, James H., Karen A. Rader. "Big Science: Price to the Present." *Osiris* 7 (1992): 3–25.

Carlsson-Hyslop, Anna. "An Anatomy of Storm Surge Science at Liverpool Tidal Institute 1919–1959: Forecasting, Practices of Calculation and Patronage." PhD diss., University of Manchester, 2010. British Library EthOS (uk.bl.ethos.538271).

Coleman, William D., Grace Skogstad. "Policy Communities and Policy Networks: A Structural Approach." In *Policy Communities and Public Policy in Canada*. Edited by William D. Coleman and Grace Skogstad. Toronto: Copp Clark Pitman, 1990.

Collins, Harry M. *Changing Order: Replications and Induction in Scientific Practice*. California: Sage, 1985.

Collins, Harry M., Trevor J. Pinch. *Frames of Meaning: The Social Construction of Extraordinary Science*. London: Routledge, 1982.

Collins, Patricia Hill. "Intersectionality's Definitional Dilemmas." *Annual Review of Sociology* 41 (2015): 1–20.

Doel, Ron E. "Scientists as Policymakers, Advisors, and Intelligence Agents: Linking Contemporary Diplomatic History with the History of Contemporary Science." In *The Historiography of Contemporary Science and Technology*. Edited by Thomas Söderqvist. Reading, UK: Harwood Academic Publishers, 1997.

Duara, Prasenjit. "The Cold War as a Historical Period: An Interpretive Essay." *The Journal of Global History* 6 (2011): 457–480.

Dylan, Huw. *Defence Intelligence and the Cold War: Britain's Joint Intelligence Bureau 1945–1964*. Oxford: Oxford University Press, 2014.

Edgeton, David. *Warfare State: Britain, 1920–1970*. Cambridge: Cambridge University Press, 2005.

Engerman, David. "Ideology and the Origins of the Cold War, 1917–1962." In *Cambridge History of the Cold War*. Edited by Melyvn P. Leffler, Odd Arne Westad. Cambridge: Cambridge University Press, 2010.

Forman, Paul. "Behind Quantum Electronics: National Security as Basis for Physical Research in the United States, 1940–1960." *Historical Studies in the Physical and Biological Sciences* 18 (1987): 149–229.

Gaddis, John Lewis. *Strategies of Containment: A Critical Appraisal of Postwar American National Security Policy*. Oxford: Oxford University Press, 1982.

Gaddis, John Lewis. *The Cold War*. London: Allen Lane, 2005.

Galison, Peter. "Bubble Chambers and the Experimental Workplace." In *Observation, Experiment, and Hypothesis in Modern Physical Science*. Edited by Peter Achinstein, Owen Hannaway. Cambridge, MA: MIT Press, 1985.

Galison, Peter. "The Many Faces of Big Science." In *Big Science: The Growth of Large-scale Research*. Edited by Peter Galison. Stanford, CA: Stanford University Press, 1992.

Geiger, Roger. "Review of *The Cold War and American Science: The Military-Industrial-Academic Complex at MIT and Stanford*." *Technology and Culture* 34 (1994): 629–631.

Giauque, Jeffrey G. *Grand Designs and Visions of Unity: The Atlantic Powers and the Reorganization of Western Europe, 1955–1963*. Chapel Hill: University of North Carolina Press, 2002.

Glaser, Charles L. "The Security Dilemma Revisited." *World Politics* 50 (1997): 171–201.

Goodman, Dena. *The Republic of Letters: A Cultural History of the French Enlightenment*. Ithaca, NY: Cornell University Press, 1994.

Goodman, Michael S. *Spying on the Nuclear Bear: Anglo-America Intelligence and the Soviet Bomb*. Redwood City, CA: Stanford University Press, 2007.

Goodman, Michael S. *The Official History of the Joint Intelligence Committee, Volume 1: From the Approach of the Second World War to the Suez Crisis*. London: Routledge, 2014.

Hamblin, Jacob D. "The Navy's 'Sophisticated' Pursuit of Science: Undersea Warfare, the Limits of Internationalism, and the Utility of Basic Research, 1945–1956." *Isis* 93, no. 1 (2002): 1–27.

Hamblin, Jacob D. *Oceanographers and the Cold War: Disciples of Marine Science*. Seattle: University of Washington, 2005.

Hankins, Thomas L. "In Defense of Biography: The Use of Biography in the History of Science." *History of Science* 17 (1979): 1–16.

Haslam, Jonathan J. *Russia's Cold War: From the October Revolution to the Fall of the Wall*. New Haven, CT: Yale University Press, 2012.

Hounshell, David. "The Cold War, RAND, and the Generation of Knowledge, 1946–62." *Historical Studies in the Physical and Biological Sciences* 27, no. 2 (1997): 237–267.

Hughes, Thomas P. "The Electrification of America: The System Builders." *Technology and Culture* 20 (1979): 124–161.

Iriye, Akira. "Historicizing the Cold War." In *The Oxford Handbook of the Cold War*. Edited by Richard H. Immerman, Petra Goedde. Oxford: Oxford University Press, 2013.

Jervis, Roger. "Was the Cold War a Security Dilemma." *Journal of Cold War Studies* 3 (2001): 36–60.

Jordan, Grant. "Sub-Governments, Policy Communities, and Networks: Refilling the Old Boots." *Journal of Theoretical Politics* 2, no. 3 (1990): 319–338.

Keuren, David K. van. "Cold War Science in Black and White: US Intelligence Gathering and Its Scientific Cover at the Naval Research Laboratory, 1948–62." *Social Studies of Science* 31, no. 2 (2001): 207–229.

Kevles, David J. "Cold War and Hot Physics: Science, Security, and the American State." *Historical Studies in the Physical Sciences* 20 (1990): 239–264.

Klijn, Erik hans. "Analysing and Managing Policy Processes in Complex Networks: A Theoretical Examination of the Concept Policy Network and Its Problems." *Administration & Society* 28, no. 1 (1996): 90–119.

Klijn, Erik hans, Joop Koppenjan. "Governance Network Theory: Past, Present, Future." *Policy & Politics* 40, no. 4 (2012): 587–606.

Krige, John. *American Hegemony and the Postwar Reconstruction of Science in Europe*. Cambridge, MA: MIT Press, 2006.

Kwon, Heonik. *The Other Cold War*. New York: Columbia University Press, 2010.

Latour, Bruno. *Science in Action: How to Follow Scientists and Engineers Through Society*. Cambridge, MA: Harvard University Press, 1987.

Latour, Bruno, Stephen Woolgar. *Laboratory Life: The Construction of Scientific Facts*. Princeton, NJ: Princeton University Press, 1986.

Laughton, Anthony, John Gould, M.J. 'Tom' Tucker, Howard Roe. *Of Seas and Ships and Scientists: The Remarkable Story of the UK's National Institute of Oceanography*. Cambridge: Lutterworth Press, 2010.

Leffler, Melvyn P., David S. Painter. *Origins of the Cold War: An International History*. London: Routledge, 2005.

Leslie, Stuart W. *The Cold War and American Science: The Military-Industrial-Academic Complex at MIT and Stanford*. New York: Columbia University Press, 1994.

Lewis, Jenny M. "The Future of Network Governance Research: Strength in Diversity and Synthesis." *Public Administration* 89 (2011): 1221–1234.

Loth, William. *Europe, Cold War and Coexistence*. London: Frank Cass, 2004.

Lundestad, Geir. *The United States and Western Europe Since 1945: From "Empire" by Invitation to Transatlantic Drift*. Oxford: Oxford University Press, 2003.

MacKenzie, Donald. "Statistical Theory and Social Interests: A Case Study." *Social Studies of Science* 8, no. 1 (1978): 35–83.

Macrajis, Kristie. "Techniphilic Hubris and Espionage Styles During the Cold War." *Isis* 101 (2010): 378–385.

McCulloch, Andrew D. *Charisma and Patronage: Reasoning with Max Weber*. Farnham, UK: Ashgate, 2014.

Menahem, Gila. "Policy Paradigms, Policy Networks and Water Policy in Israel." *Journal of Public Policy* 18, no. 3 (1998): 283–310.

Merton, Robert K. "Priorities in Scientific Discovery: A Chapter in the Sociology of Science." *American Sociological Review* 22 (1957): 635–659.

Moore, Kelly. *Disrupting Science: Social Movements, American Scientists, and the Politics of the Military, 1945–1975*. Princeton, NJ: Princeton University Press, 2008.

Morrell, Jack, Arnold Thackray. *Gentlemen of Science: Early Years of the British Association for the Advancement of Science*. Oxford: Oxford University Press, 1981.

Mukerji, Chandra. *A Fragile Power: Scientists and the State*. Princeton, NJ: Princeton University Press, 1989.

Nye, Mary J. *Blackett: Physics, War, and Politics in the 20th Century*. Cambridge, MA: Harvard University Press, 2004.

Nye, Mary J. "Scientific Biography: History of Science by Another Means." *Isis* 97 (2006): 322–329.

Nye, Mary J. *Michael Polanyi and His Generation: Origins of the Social Construction of Science*. Chicago: University of Chicago Press, 2011.

Oikonomou, Alexandros-Panagiotis. "The Hidden Persuaders: Government Scientists and Defence in Post-war Britain." PhD diss., Imperial College London, 2011. British Library EthOS (uk.bl.ethos.537242).

Oreskes, Naomi. "A Context of Motivation: US Navy Oceanographic Research and the Discovery of Sea-Floor Hydrothermal Vents." *Social Studies of Science* 33 (2003): 697–742.

Oreskes, Naomi, Ronald Rainger. "Science and Security Before the Atomic Bomb: The Loyalty Case of Harald U. Sverdrup." *Studies in History and Philosophy of Science Part B: Studies in History and Philosophy of Modern Physics* 31 (2000): 309–369.

Pestre, Dominique. "Science, Political Power and the State." In *Companion to Science in the Twentieth Century*. Edited by John Krige, Dominique Pestre. London: Routledge, 2003.

Pestre, Dominique. "Debates in Transnational and Science Studies: A Defence and Illustration of the Virtues of Intellectual Tolerance." *British Journal for the History of Science* 45, no. 3 (2012): 425–442.

Pickering, Andrew. *Constructing Quarks: A Sociological History of Particle Physics*. Edinburgh: Edinburgh University Press, 1984.

Pinch, Trevor J. "Sociology of Science." In *Reader's Guide to the History of Science*. Edited by Arne Hessenbruch. London: Fitzroy Dearborn Publishers, 2000.

Porter, Theodore M. "Is the Life of the Scientists a Scientific Unit?." *Isis* 97 (2006): 314–321.

Pyenson, Lewis. "An End to National Science: The Meaning and the Extension of Local Knowledge." *History of Science* 40 (2002): 251–290.

Rainger, Roland. "Patronage and Science: Roger Revelle, The U.S. Navy and Oceanography at the Scripps Institution." *Earth Sciences History* 19 (2000a): 58–89.

Rainger, Roland. "Science at the Crossroads: The Navy, Bikini Atoll, and American Oceanography in the 1940s." *Historical Studies in the Physical and Biological Studies* 30 (2000b): 349–371.

Richards, Joan L. "Introduction: Fragmented Lives." *Isis* 97 (2006): 302–305.

Roberts, Peder. "Intelligence and Internationalism: The Cold War Career of Anton Bruun." *Centaurus* 55 (2013): 243–263.

Rozwadowski, Helen. *The Sea Knows No Boundaries: A Century of Marine Science Under ICES*. Seattle: University of Washington Press, 2004.

Saunier, Pierre-Yves. "Entry: Transnational." In *The Palgrave Dictionary of Transnational History*. Edited by Akira Iriye, Pierre-Yves Suanier. New York: Palgrave Macmillan, 2009.

Shapin, Steven, "The Politics of Observation: Cerebral Anatomy and Social Interests in the Edinburgh Phrenology Disputes." In *On the Margins of Science: The Social Construction of Rejected Knowledge*. Edited by Roy Wallis. Keele, UK: Keele University Press, 1979: 139–178.

Shapin, Steven, Simon Schaffer. *Leviathan and the Air Pump: Hobbes, Boyle, and the Experimental Life*. Princeton, NJ: Princeton University Press, 1985.

Shibusawa, Naoko. "Ideology, Culture, and the Cold War." In *The Oxford Handbook of the Cold War.* Edited by Richard H. Immerman, Petra Goedde. Oxford: Oxford University Press, 2013.
Smith, Crosbie, M. Norton Wise. *Energy & Empire: A Biographical Study of Lord Kelvin.* Cambridge: Cambridge University Press, 1989.
Terrall, Mary. "Biography as Cultural History of Science." *Isis* 97 (2006): 306–313.
Thorpe, Charles. *Oppenheimer: The Tragic Intellect.* Chicago: University of Chicago Press, 2006.
Thorpe, Charles, Steven Shapin. "Who Was J. Robert Oppenheimer? Charisma and Complex Organization." *Social Studies of Science* 30, no. 4 (2000): 545–590.
Trachtenberg, Marc. *A Constructed Peace: The making of the European Settlement, 1945–1963.* Princeton, NJ: Princeton University Press, 1999.
Turchetti, Simone. *The Pontecorvo Affair: A Cold War Defection and Nuclear Physics.* Chicago: University of Chicago Press, 2012.
Turchetti, Simone, Peder Roberts. *The Surveillance Imperative: GeoSciences During the Cold War and Beyond.* London: Palgrave Macmillan, 2014.
Vanhaute, Eric E. "Who Is Afraid of Global History? Ambitions, Pitfalls and Limits of Learning Global History." *Österreichische Zeitschrift für Geschichtswissenschaften* 20, no. 2 (2009): 22–39.
Waarden, Frans van. "Dimensions and Types of Policy Networks." *European Journal of Political Research* 21, no. 1 (1992): 29–52.
Walker, Mark. "The 'National' in International and Transnational Science." *British Journal for the History of Science* 45 (2012): 359–376.
Warner, Geoffrey. "The Geopolitics and the Cold War." In *The Oxford Handbook of the Cold War.* Edited by Richard H. Immerman, Petra Goedde. Oxford: Oxford University Press, 2013.
Weir, Gary E. *An Ocean in Common: American Naval Officers, Scientists, and the Ocean Environment.* College Station, TX: Texas A&M University Press, 2001.
Westad, Odd Arne. *The Global Cold War: Third World Interventions and the Making of Our Times.* Cambridge: Cambridge University Press, 2006.
Westad, Odd Arne. "The Cold War and the International History of the Twentieth Century." In *The Cambridge History of the Cold War.* Melvyn P. Leffler, Odd Arne Westad. Cambridge: Cambridge University Press, 2012.
Williams, L. Pearce. "The Life of Science and Scientific Lives." *Physis* 28 (1991): 199–213.
Ziegler, Charles A., David Jacobson. *Spying Without Spies: Origins of America's Nuclear Surveillance System.* Westport, CT: Greenwood Publishing, 1995.

CHAPTER 2

Oceanographers at War

Just over a month after VE Day, British oceanographer James Carruthers boarded the German research ship *Borgen* and interrogated his pre-war scientific colleagues. An obituary written by the Director of the Deutsches Hydrographisches Institut, Günther Böhnecke, described Carruthers' arrival in Flensburg:

> How great was our surprise therefore when on 14 June 1945 Dr Carruthers turned up in Flensburg [Germany] on board our research ship Borgen which was then the home of the permanent staff and most of the instruments of our evacuated marine laboratory and, on behalf of the occupying powers, proceeded to interrogate us about the work of the laboratory and the results obtained since 1939.[1]

This anecdote reminds us of the new roles that ocean science and British oceanographers played during and after the conflict. The Second World War reshaped and reframed the relationships between scientists, the Admiralty, and other key structures within the British state. These reconfigured relationships aided in the establishment of a new policy network, facilitating the growth of oceanographic studies which were highly significant to the operations of the Royal Navy. The war years thus proved decisive in the establishment of a "military" oceanography and shaped the Cold War research trajectories of British oceanographers.

S. A. Robinson, *Ocean Science and the British Cold War State*,
Palgrave Studies in the History of Science and Technology,
https://doi.org/10.1007/978-3-319-73096-7_2

In order to highlight the lasting impact of the Second World War on ocean science in Britain this chapter focuses on three British oceanographers: George Deacon (Admiralty Anti-Submarine Establishment, HMS *Osprey*), Edward Bullard (Admiralty Mine Design Department), and James Carruthers (Hydrographic Office). All three worked on military-related subjects and began the war employed in civilian capacities; none had served before in a military research establishment. Yet their wartime research on sonar (known in Britain by the acronym ASDIC),[2] mining, and hydrography respectively made significant contributions to the war effort and Britain's post-war defensive and geopolitical polices. At the outset of the conflict each of these oceanographers entered naval service at the lowest level; by its end they were in a position to contribute to post-war military science policy making. Of the three, George Deacon was the most successful in establishing new connections between networks within institutions. This helped him to gain a position of influence which would prove vital in the construction of a new policy network furthering military oceanography.

The UK has a long history of sponsoring oceanographic studies, which are traditionally associated with the mapping of the oceans for the Royal Navy, and tasks involved in the control, protection, and economic exploitation of areas of political and commercial interest to the British Empire. The traditional home of physical oceanography within the Royal Navy was the Hydrographic Office, which was established in 1795 and served as the original scientific department of the Royal Navy. The office, overseen by the Hydrographer, was tasked with producing charts, marking sunken ships and identifying the location of minefields. The Royal Navy also commissioned studies of the deep sea, supporting civilian men of science, such as in the case of the famous *Challenger* expedition (1872–76).[3]

In the first half of the twentieth century sea studies proliferated in Britain. This was partly because of their importance to defence tasks in the First World War and the organisation of international collaborative projects. In 1920 William Herdman, an oceanographer and president of the British Association for the Advancement of Science (BAAS), delivered a presidential address calling for a new *Challenger*-style expedition, which was unsuccessful due to a lack of funds.[4] This call did, however, result in the creation of the Discovery Committee, a body that was intended to uphold British research traditions in deep-sea ocean exploration funded from taxes on the whaling industry of the Southern Ocean. The committee had the imperial and geopolitical aim of "developing" the Falkland Island Dependencies (a British territory) in the South Atlantic.[5] The

Discovery Committee used this funding to build two vessels: RRS *William Scoresby* (1925) and RRS *Discovery II* (1929).[6] In addition to addressing the Navy's requests for assistance and surveillance of distant places, the committee's main task was to provide information on uncharted oceanic waters and their characteristics (sea currents especially) in order to facilitate the work of British whalers and other industrialists working in the fisheries sector.

Before the Second World War, Britain's efforts to chart the ocean were consistent with its imperial ambitions around the UK, in the South Atlantic, the Indian Ocean, and in some areas of the Pacific. This global project was centralised in the office of the Hydrographer, with other institutions such as the Discovery Committee only assisting by association through the publication of scientific papers and by exposing the naval crew and officers of their research vessels to the Southern Ocean. Other smaller oceanographic ventures within Britain tended to focus on inshore or home waters. The Ministry of Agriculture, Fisheries and Food (MAFF) became responsible for British involvement in the activities of the International Council for the Exploration of the Seas (ICES), founded in 1902, through its Fisheries Research Laboratory at Lowestoft, established in 1919.[7] Here joint Anglo-German and Dutch experiments into currents in the southern North Sea were undertaken. Theoretical research on tides and waves was centred on the Tidal Institute in Liverpool (established in 1919), which was closely tied to the university, and in Cambridge, as a part of the work of the Department of Geodesy and Geophysics (established in 1931). The Tidal Institute used mathematical methods to provide information concerning tides to the merchant and naval marine.[8] Work at the University of Cambridge was closely related to the study of the physical properties of the earth, particularly geomagnetic research.

A significant volume of data and new scientific papers were produced with the growing number of ocean scientists, but this was only one part of the expansion of the discipline in Britain, which entailed gathering knowledge on distant places, at least as far as Britain's sphere of influence was concerned. In the years prior to 1939 oceanography already existed in several separate research contexts receiving UK government funds, and it was the need for military readiness in naval operations that increased the budget of those specific branches of oceanographic research that had significance for these operations.

In forging these plans, British naval commanders were primarily trying to catch up with the work commissioned by other navies to oceanographers active in rival nations. While these British military and industrial sponsors wished to cover other oceans selectively, foreign patrons began to see the advantages of global mapping of the seas. One of the earliest attempts to forge a global concept of oceanography was a report commissioned by the US National Academy of Sciences in 1927. The report sought to 'consider the share of the United States of America in a worldwide program of oceanographic research'.[9] Authored by Scripps Oceanographic Institute director, Thomas Wayland Vaughan,[10] the report was eventually published in 1937 under the title "International Aspects of Oceanography". Vaughan attempted to collate data and oceanographic knowledge, and through the extensive use of charts the report described what was known about the world's oceans, and which nation or expedition had contributed this knowledge. The report also provided a complete list of all of the research centres across the globe with an interest in oceanographic study. Vaughan travelled extensively[11] gathering material for his compilation of oceanographic research undertaken globally up to 1936. This was the first global study of the physical characteristics of seas and oceans (temperature, salinity, pressure [depth], currents) as opposed to broader approaches that also included marine biology. This separation into physical and biological spheres, previously avoided, reflected the emerging military involvement with physical oceanography studies, particularly those exploring underwater sound and echo location. Scientists such as George Deacon who had little interest in fisheries science also sought to shift focus away from biological priorities towards physical by courting military-scientific objectives.[12]

By the late 1930s the value of physical oceanographic knowledge was becoming apparent to the hydrographic service of the Royal Navy. Vaughan's report reflected interest at the highest scientific—and, by association, government—levels in the USA in quantifying knowledge of oceanic characteristics, so as to both keep abreast of research being undertaken elsewhere, and to identify where the United States was in relation to other countries' programmes. Placing a catalogue of research into one monograph, as Vaughan had done, allowed researchers to quickly identify others' work, what data might be available, and most significantly, for which areas of the ocean.[13]

Furthermore, by 1939 oceanography had become established as an independent scientific discipline of global significance. Papers were published in many languages, although many nations' research focus tended to be local or regional. The specific geopolitical or imperial functions of prominent studies meant that wider knowledge of the oceans outside a particular nation's area of interest was lacking. However, high-profile round-the-world expeditions, such as the German *Meteor* (1925–27) and the Danish *DANA II* (1928–30), which did not specifically focus on national waters and were primarily a geopolitical flag-waving exercise, gathered large amounts of data in marine geophysics and marine biology respectively.[14]

While Vaughan's effort had uncovered the power of oceanographic studies for one nation, during the inter-war period it was becoming evident that one nation alone could not gather a complete set of oceanographic knowledge ensuring global coverage. This understanding instigated more international collaborative efforts and the sharing of data, especially between the oceanographers of allied nations such as Britain, America, and the Scandinavian countries.

In the late 1930s, the growing potential menace to military and commercial shipping across the Atlantic in any future conflict posed by Nazi Germany's renewed submarine construction programme focused the attention of British military planners on the need to obtain more, and more accurate, oceanographic data. As war seemed ever more likely, Royal Navy officials agreed to survey for themselves developments in physical oceanography occurring in the USA and elsewhere and fund new studies under the aegis of the Hydrographic Office. New tasks comprised the collection of new data through exploration, as well as the analysis of data already collected in existing British and foreign publications, including Vaughan's pioneering study. Additionally, the Royal Navy Hydrography Office and the Office of the Oceanographer of the US Navy began sharing data and working directly with physical oceanographers, creating a relationship that would endure into the Cold War. We shall now see that three oceanographers gained considerably from this process, J.N. Carruthers, Edward Bullard, and George Deacon. Ultimately Deacon would employ the networks forged during these years to rise through the ranks and become the leading figure in British military oceanography.

The Hydrographic Office and the Use of Oceanographic Knowledge in Wartime

Before the outbreak of the Second World War, the Royal Navy Hydrographer, Vice Admiral Sir John Edgell, was consulted by naval officers on issues related to forecasting operational conditions which were of increasing significance to those planning for wartime sea operations. Edgell, who had been the Hydrographer since 1932, had strengthened and rebuilt the surveying service and was the first Hydrographer to be elected a Fellow of the Royal Society in over a hundred years.[15] His election was based not on his scientific work (he never published a peer-reviewed scientific paper) but on the administrative work he had done to expand the hydrographic service in particular and to reinvigorate a scientifically minded Navy in general.[16] To cope with the influx of requests Edgell requested in 1937 that James Carruthers of the MAFF Fisheries Research Laboratory (Lowestoft, Suffolk) be seconded to his department.[17]

Carruthers had served in the infantry during the First World War, after which he studied geology at the University of Leeds, graduating in 1920. Subsequently he was employed as a hydrologist at the Fisheries Research Laboratory studying the effects of currents and drift in the North Sea. Carruthers, a "veritable bookworm", was considered exceptional amongst British oceanographers. He spoke fluent German, but had never previously been responsible for studies associated with naval operations.[18] In 1937, however, his comprehensive bibliographical knowledge of British and international oceanography was exactly what Edgell was looking for. Carruthers had accrued significant international experience whilst working for the ICES southern North Sea committee during the 1920s. His interaction with non-British ocean science provided him with both an international network of fellow oceanographers and knowledge of research undertaken in other nations, making him a valuable resource for both data collection and dissemination within the British community.

Carruthers' role at the Hydrographic Office was to produce qualified reports and memoranda for use in naval operations. The Admiralty was becoming increasingly concerned that information about the ocean spaces in which naval operations took place was scant, and that as naval activities became more complex, a more refined understanding of sea characteristics (temperature, currents, beach hydrography, and likely climatic conditions and sea states) was needed. After the conflict, Carruthers justified the Admiralty's need for oceanographers in a single sentence: 'The efficacy of many modern operations and of many modern sea weapons, must depend upon an adequate knowledge of the temporal and spatial vagaries of the medium upon and within which they are to be carried out or employed.'[19]

Carruthers' argument that the Admiralty needed to invest more funds in the activities of oceanographers, and that the Royal Navy should make greater use of oceanographic knowledge, was based on his wartime experiences. He had spent the war convincing naval commanders to allow him and his colleagues to convert basic sea research to answer applied problems faced by the Admiralty laboratories. Working up existing knowledge into simple memos and supplying charts, photographs and diagrams was a straightforward method of providing detailed answers to the oceanographic questions posed by Admiralty scientists and officers who lacked access to a scientific library.[20] In this role Carruthers could be said to have been the first scientific middleman connecting oceanographic research with the practical development work of Admiralty research establishments.[21]

Carruthers, however, had done much more than simply work up existing knowledge. His first task at the Hydrographic Office had been to write a report on the practices he expected German marine scientists to have suggested to their navy.[22] This kind of basic open-source intelligence became increasingly useful during the conflict. From the starting point of this initial report, the questions Edgell asked Carruthers to address became much broader, although the majority concerned German submarine operations. Carruthers surveyed the positions of mines and provided information on the likely courses of enemy vessels and drifting mines, the currents of the North Sea, the currents and densities of seawater in the Dardanelles and Bosporus, and the physical properties of the Strait of Gibraltar.[23] To do so he 'work[ed] up information of Danish, Norwegian, German and Portuguese origin'.[24]

All of this work was intended to show the usefulness of scientific expertise to modern conflict and as a result, by the end of the war, Carruthers' influence within the Admiralty had grown considerably.[25] He had his own team within the Hydrographic Office and was consulted when preparations were being made for key naval operations. In preparation for D-Day in June 1944, his assistants reviewed literature on the French coastline and prepared nearly one hundred documents on French beaches, before beginning an examination of Russian oceanographic literature. Deacon recalled in Carruthers' obituary: 'He worked very long hours: everyone who sought his help will remember the large handwritten manuscripts, photocopies, maps, diagrams and references that came back by return post.'[26] After the war Carruthers, fearing that the Navy would forget his efforts, stated that 'it cannot be ignored that the services of persons having a knowledge of (and in possession of) the very diverse and complicated literature pertaining to the subject should be available'.[27] Seeking to capitalise on his contribution by stating clearly the military implication of discontinuing his work, he argued: 'Many mines put out for a specific purpose in Norwegian waters during the last War, must have been useless for the purposes intended owing to non-consideration of the very complicated water density conditions.'[28] If widely read physical oceanographers were not consulted, Carruthers argued, avoidable failures would occur, which went against military notions of planning and preparedness.[29]

In preparing new reports, Carruthers was—from his perspective—simply re-presenting research and data from Norwegian scientific papers that dealt with fishery problems and currents research that proved useful in addressing the difficulties facing the Admiralty. Yet, this work involved translation into English of complex foreign scientific papers and research conducted in distinct disciplinary and institutional contexts. Carruthers' translation showed that research conducted on and in the seas could be applied to multiple national and transnational issues, but also placed the Norwegian research into a specifically British milieu, opening it up to new applications. Translation of existing research had a broad impact within the Royal Navy scientific war effort in areas including ocean conditions and their impact on underwater acoustics for anti-submarine warfare, ocean current science to recover downed pilots in the North Sea, and French beach profile information in the planning for various amphibious landings.

The employment of oceanographers like Carruthers in the Hydrographic Office shows that research not directly connected to defence could have direct strategic relevance in time of war.[30] Compiling existing research into accessible reports reduced the amount of effort expended on basic research in wartime at Royal Navy research stations.[31] Additionally, scientists were able to gain precious intelligence by surveying the literature and suggesting how the enemy might use its ocean scientists. They understood, for instance, the specific specialities of German oceanographers and their strengths in pre-war research.

Whilst Carruthers became a central figure in the redistribution and supply of information to both Admiralty planners and scientists working within research establishments, he did not in the end play a key role in British oceanographic affairs. However, while Carruthers was busy translating and "working up" research produced by foreign scientists, his colleague and friend George Deacon was working on original anti-submarine warfare research in support of the Royal Navy's fight in the Battle of the Atlantic.

Military Utility of Oceanography: ASDIC and the Submarine Menace (1940–43)

George Deacon had had little interaction with the military prior to the war. Born in 1906 in Leicester, in the English Midlands far from the sea, he studied chemistry at Kings College London, graduating in 1926 with a first-class honours degree and an ambition to become a schoolteacher. During 1927 he replied to an advert he saw in the newspaper for a marine chemist to join the Discovery Committee investigations. Deacon was appointed and posted to the RSS *William Scoresby*. At the end of the *Scoresby* expedition, Deacon was sent almost immediately to the new Discovery Committee vessel the RSS *Discovery II* in Cape Town to increase the scientific complement. He stayed for the ship's second expedition and by the fourth voyage (1935–37) was appointed the principal scientist. His scientific account of the second expedition, "The hydrology of the Southern Ocean", was approved by the University of London for a DSc award, and on the basis of this work Deacon was elected FRS in 1944. However, in 1939, having failed to be selected for the latest *Discovery* expedition, he wrote to the Director of Scientific Research at the Admiralty, Charles Seymour Wright, asking if there was any way in which he could help the war effort (Fig. 2.1).

Upon entering the Royal Navy's Division of Scientific Research, Deacon was appointed to the Anti-Submarine Establishment Portland (HMS *Osprey*) to assist with work on the challenges posed by the deepening battle between German submarines and British merchant vessels in the North Atlantic. He remained a civilian, not a commissioned military officer. In January 1940, undertaking research on board HMS *Kingfisher,* he was made to carry a card which clearly stated that he had been 'specially instructed that he is <u>not</u> to participate in the fighting of the ship or to exercise while on board, any function different from that in which he was embarked, except such humanitarian functions as the succouring of the sick and wounded'.[32] Despite his limited prior experience of the work of the Royal Navy scientific establishments he was expected to quickly assimilate into a new network of military-scientific researchers.

The need to recruit and rapidly deploy scientists into the Royal Navy's research structure reflected the challenge of ensuring that Britain was not cut off from vital war supplies. This made the North Atlantic one of the most scientifically researched theatres of the Second World War. During the war Great Britain was entirely dependent on its large merchant navy for supplies of food, goods, and certain materials. Wartime needs merely exacerbated

Fig. 2.1 George Deacon on the deck of the RRS *Discovery II* on a Discovery Investigations Committee Voyage, late 1930s. (Image from the Archives of the National Oceanographic Library, National Oceanography Centre. Southampton)

the country's existing reliance on imported commodities, as had occurred to a lesser extent during the First World War.[33] Patrick Blackett's fledgling field of Operational Research grew out of work on convoy systems and corresponding merchant fleet survival rates.[34] Others such as Deacon worked directly on anti-submarine warfare (ASW) weapons and tools. Germany's unrestricted submarine warfare campaign had caused significant damage to British shipping in 1917, leading the Admiralty to assume that the Germans would follow a similar strategy during the early years of the Second World War.[35] When the British Empire became the sole power fighting Germany between the summer of 1940 and the summer of 1941, it became heavily reliant on supplies from the US, making the submarine threat even greater.[36] The Royal Navy was relatively successful at destroying German surface warships—as demonstrated by the sinking of the battleship *Bismarck* in 1941. However, Germany's submarines were seen as the chief threat at sea because they were much more difficult to track and destroy before first sinking a vessel and thereby revealing themselves.[37] With the submarine threat foreseen, inter-war research at the Admiralty Research Laboratory had been directed towards anti-submarine warfare.[38]

A key technology developed before the war was echo-sound location devices, a system known as active sonar.[39] This was more accurate than the hydrophones used during the First World War, which had worked passively, relying on listening for sounds in the ocean. Active sonar systems, by contrast, worked in much the same way as radar, emitting pulses (sound waves) that bounced off targets in the water. Any returning sound was registered by hydrophones on the ship, allowing for the tracking of underwater objects. However, the emission of the pulse also gave away the position of the emitting vessel.[40] Prior to the Second World War, and throughout the conflict, Britain and America collaborated closely on ASW, especially sonar technologies, and on underwater acoustics.[41]

Sonar detection of submarines was affected by variations in ocean conditions. Oceanographers in the United States had identified the thermocline, which obstructed the effectiveness of sonar. A thermocline is a distinct layer in a large body of water whose temperature will be different to that above and below. As sound passes through the thermocline, the sound wave is bent like light in a prism by the temperature, salinity, and pressure of the water it passes through. This results in blind zones, where the thermocline causes the sound waves to be curved around a water mass in which a submarine can intentionally hide, or unintentionally remain beyond the detection of pursuers. Hamblin describes this elucidation of the properties of this phenomenon as 'the ultimate research agenda for both scientists and the military…scientists could map the locations of thermoclines for the Navy's operational use without impeding whatever other research they prioritized for themselves'.[42]

Following the bombing of the south coast of England during the late summer of 1940, the Royal Navy was forced to disperse and evacuate its R&D facilities from its traditional bases at Portsmouth, Plymouth, and Portland. This involved moving personnel and equipment across the country to any location where suitable facilities could be found. In the case of ASW research this meant relocating to William Fife's yacht-building yard, located on the Ayrshire coast at Fairlie, Scotland.[43] The re-location gave greater freedom of operation for HMS *Osprey*. The close proximity of skilled craftsmen and the calm waters of the Firth of Clyde provided a test range with a great deal of independence from the shared facilities and resources at Portland. This allowed for the permanent seabed installation of hydrophones linked back to shore-based facilities, creating an underwater acoustics test range. With a fixed range, experiments could be carried out regularly; not only could new technologies be tested, but the science of the hydrology of the range waters could be investigated. The wartime *Osprey* was equipped to study underwater acoustics as well as other scientific phenomena and new detection equipment.

Deacon began to benefit from exchanges with foreign colleagues from allied nations. For instance, in addition to the British scientists working at HMS *Osprey*, the fall of Norway brought Norwegian evacuee experts to join British research establishments. Norwegian electrical engineer Fredrik Møller headed up a team that included Henrik Nødtrvedt, Ingvald Engelsen, Ole Harbek, and Torvald Gerhardsen,[44] contributing expertise from the Bergen Geophysical Institute, which had conducted numerous studies of the hydrographic conditions of the waters off the coast of Norway, beginning in 1909. After the war and into the Cold War these Anglo-Norwegian connections were very important to Deacon's network of international contacts, especially—as we shall see—in the context of NATO-sponsored oceanographic work.[45] Very little material about the day-to-day running of HMS *Osprey* exists, but numerous reports were written dealing with such topics as anti-submarine warfare, sonar development, and oceanographical characteristics of the Atlantic.[46] They reveal the importance of Deacon's work during this period and its contribution to the eventual deployment of a new instrument, the bathythermograph, on board Royal Navy submarines.

Oceanography at the Admiralty Research Establishments

American oceanographers had discovered during trials in the Straits of Florida that the thermocline affected sonar performance. These trials had been conducted using bathythermographs. A bathythermograph (also known as a BT) was used to simultaneously measure ocean temperature and depth, and to plot them with a stylus on an inexpensive smoked glass slide that could be easily removed from the instrument and read against a calibration chart. The instrument was lowered from the back of either a stationary or moving surface or submarine vessel, and measurements were continuously taken throughout the instrument's ascent. Its development in the late 1930s enabled, in principle, a synoptic view of the world's oceans, although the first large-scale deployment was on board US submarines operating in the Pacific theatre whose crew needed to understand local conditions.[47] It was only after the war, when many hundreds of these instruments had been constructed for war work, that they were deployed on a scale that could potentially enable oceanographers to understand global changes in the characteristics of water temperature and deep ocean currents. By 1943 reports from submarine commanders led scientists at *Osprey* to believe that problems identified in the refraction of sound waves were hindering the effectiveness of close-range detection and the tracking of submarines, reducing the effectiveness of new surface-deployed anti-submarine weapons, such as the new styles of depth charges being fitted to escort vessels. In discussions between scientists in the USA and Britain in 1940–41, American naval scientists had concluded that the British were only interested in long-range detection of submarines where the issue of the thermocline was less marked.[48] George Deacon and another *Osprey* scientist, Henry Wood, were tasked with studying this problem.[49]

In order to conduct British trials, Wood was sent with American-designed bathythermographs on a convoy to northern Russia in October 1943.[50] The dangerous North Russia convoy route was chosen because in the winter of 1942–43 an escort group had failed to make a single contact with an enemy submarine, despite being attacked several times. Long-range echoes were obtained, but these were subsequently lost when the range was closed.[51] Wood took three of the four British-owned bathythermographs with him, but two were lost early on because of a lack of knowledge about what strength of wire was required to lower them. Wood was therefore very careful with his one remaining instrument and made very few actual measurements. Ultimately Wood could not explain what sonar conditions had befallen the previous convoy, but was able to suggest that the culprit could be the "layering" effect caused by the mixing of the Atlantic waters with high salinity densities and the cold Polar waters with relatively low salinity densities, which indicated the presence of thermocline activity. In such conditions, Wood concluded that while long-range contacts were achievable, contacts with vessels below the boundary between the two waters was likely to be difficult, if not impossible.[52]

Deacon composed a preliminary report, supposedly co-authored by himself and Wood, on the effect of temperature gradients.[53] Wood probably had little say in the wording of the report, however, as he was away at sea and its conclusions were based on US research rather than his own experiments in the Arctic convoys which were still underway. The report included a chart of the annual range of temperature differences at the surface of the eastern North Atlantic Ocean. The data for this came from Bjørn Helland-Hansen's 1932 analysis of the average monthly gradients in the area, produced at the Geophysical Institute in Bergen. It was presumably the interaction with his Norwegian colleagues that had made him aware of this literature and data.[54] Data and previous research was also shared between British and Scandinavian meteorologists, with expertise developed at Bergen transferred via Sverre Petterssen to the British Admiralty and Meteorological Office which assisted with the weather forecasting for the D-Day landings in June 1944.[55] This research showed that it was impossible to compensate or modify the sonar equipment permanently. However, daily adjustments could be made if salinity and temperature could be measured in the area of deployment by the ship operating the sonar equipment.[56] Beyond this, only general recommendations could be given on the conditions that sonar operators could expect to encounter. A note attached to Wood's report by the captain of the Third Destroyer Flotilla shows that the Royal Navy was prepared to undertake scientific measurements if it meant they could successfully detect and counter German submarines.[57]

The battle with German submarines in the North Atlantic was becoming the Royal Navy's primary mission, as thousands of tons of munitions, men, and material were being imported into Britain in preparation for an eventual invasion of France. That Deacon was able to play even a small part in this scientific battle against the "greatest" naval threat to Britain was a personal advantage; it brought his work to the attention of senior naval commanders, and showed the value of scientific studies of the oceans and their potential role in improving detection technologies.

The North Atlantic was not the only front on which German submarines were exacting heavy losses on British shipping.[58] During 1943 the North African front closed and the Allies turned their attention north with an amphibious assault on Sicily.[59] The fourth British bathythermograph was dispatched to HMS *Templar*, a submarine operating in the Mediterranean during the summer of 1943.[60] In deploying a bathythermograph on board a submarine, the Royal Navy was not just attempting to seek and destroy enemy submarines. In the Mediterranean British submarines were also at risk from German and Italian anti-submarine craft; the submariners were seeking not to overcome the refraction problem, but to take advantage of it by using the "hidden" zone to hide from Axis anti-submarine patrols. The US Navy was undertaking similar trials in the Pacific, with most US submarines being issued with a bathythermograph and instructions to take daily readings to work out at what depth they would become invisible to enemy sonar.[61] No scientist accompanied the

Templar, unlike the northern convoy experiment, but Admiralty commanders deemed it to have demonstrated promise. Just six weeks after the cruise the Commander-in-Chief of the Royal Navy in the Mediterranean approved the use of the bathythermograph on board the vessels under his command.[62] However, fitting these instruments to submarines required the approval of the Admiralty, so matters were passed back up the chain of command to Whitehall. With no official oceanographic programme in either the North Atlantic or the Mediterranean, oceanographers had nevertheless proved that their knowledge had strategic value that could be utilised by combat forces at sea. Oceanographic science had proved useful in the war effort, and Royal Navy commanders intended to use oceanographic instruments more frequently for their anti-submarine vessels.

In September 1943 the Admiral commanding the Royal Navy Submarine Fleet requested that fifty bathythermographs be obtained from the United States and fitted into British submarines.[63] Under the lend-lease scheme this request could be easily granted, and by March 1944 the equipment had been dispatched to the United Kingdom. The Admiral instructed them to be fitted in all the Navy's most advanced vessels operating in the warmer waters of the Mediterranean and the Far East, where it was perceived that 'oceanographic conditions will favour its use'.[64] This statement conveys the Admiral's understanding of the oceanographic principles underlying detection, and indicates the extent to which ocean science was beginning to be of use to the Royal Navy. Data taken from bathythermographs would be used to make daily adjustments to ASDIC sets, thereby improving detection of enemy submarines by indicating exactly at what depths the thermocline made them invisible to sonar.

Along with the bathythermograph, each British submarine was issued with two texts, "Instructions for the Installation, Care and Use of Submarine Bathythermograph" and "Best Depth of Escape for Submarines".[65] These had been prepared in large part by Deacon, based on his work at HMS *Osprey*. Deacon's report "Use of the Bathythermograph in British Submarines" showed the extent to which he had come to understand the operational benefit of ASDIC and the way the Navy operated. Deacon outlined plainly the potential benefits to be derived from the detection technique: 'a submarine which dives below a sharp density layer hears the ASDIC transmissions and propeller noise of a surface vessel with reduced efficiency, and she gains some immunity from echo and listening detection at long ranges'.[66] The technique allowed Deacon to get close to Navy officials and understand key aspects of the strategies they were adopting to counter the menace of submarine warfare. In turn this gave him the opportunity to consider how to increase his influence, especially as the urgency of setting up a group or a centre devoted to military aspects of oceanography was becoming more apparent to naval commanders.

Deacon also gained influence in the upper echelons of Navy scientific circles and enabled detection techniques to be further refined by establishing relations in the USA[67] which he used to reinforce his position at home and eventually to become the point of contact between military-related sea studies in Britain and those in the USA. Reports from HMS *Osprey* were often shared with experts in the US Navy who provided comments and details of their own findings. In May 1941 the Admiralty established the British Admiralty Delegation in Washington to ensure anti-submarine warfare science and technology cooperation. This body arranged numerous visits by staff from Admiralty research establishments to work in the US Navy dockyard to share knowledge and information regarding ASDIC.[68] Cooperation on sonar was the largest collaborative international project undertaken by oceanographers during the Second World War.

In December 1943 Deacon travelled to the United States to meet fellow military oceanographers at institutions on both the East and West coasts as well as the Navy Department in Washington, D.C.[69] The fact that he was allowed to undertake the dangerous voyage across the Atlantic at this time indicates that a great deal of importance was beginning to be attached to links with US oceanography by the Admiralty, particularly by the Hydrographer, Sir John Edgell. On arrival it was obvious to Deacon that 'the facilities provided by the U.S. Authorities afforded unique opportunities for the discussion of each aspect of the work with those actively engaged in it, and made it easy to appreciate the rapid development of the subject'.[70] By this time the Battle of the Atlantic had already turned. The Germans lost fifteen submarines in April 1943 and a further thirty-three in the first twenty-four days of May.[71] Improved use of the ASDIC sets, together with increased air coverage from bases in Iceland and the UK, was proving decisive in defeating German naval forces. British submarines, along with those of the US Navy, were able to use the bathythermograph to improve their own protection from Japanese and German ASW techniques.

Consequently, Deacon began to look for a new challenge. In the summer of 1944 this would lead to the establishment of the Oceanographic Section at the Admiralty Research Laboratory at Teddington (ARL). By then Deacon had established himself at the centre of a number of networks concerning the military applications of oceanography, which united Navy personnel and planners with oceanographers both in Britain and the USA. These connections would make him a suitable candidate, together with Carruthers, to carry forward fundamental research which would also be of interest to the military at the end of the war. The geophysicist Edward Bullard was also by then providing a vital contribution to naval war work.

Disharmony in Military-Scientific Relationships

Whilst Deacon and Carruthers clearly developed their military connections during the war, other scientists found their wartime experience frustrating. Oceanographers who had previously held university posts found working inside the wartime military machine somewhat difficult. This was particularly true of the scientists working at the Cambridge Department for Geodesy and Geophysics, who despite forging a working relationship with the Royal Navy before the war, never quite settled within the Admiralty research structure. The case of Edward Bullard contrasts with the narrative of essentially harmonious network building between oceanographers and military officers in the Hydrographic Office and in certain Admiralty research establishments. It is not surprising that scientists who found life within the military machine tolerable were more likely to remain within its structures after the end of the war, but it is important to remember that the careers of those who later remembered the war as a frustrating time were also shaped by these early experiences. These experiences go some way to explaining why Deacon was better placed than Bullard to direct the study of military oceanography in the UK after the war.

Edward Bullard obtained a first-class honours degree from Cambridge in Natural Sciences before entering the Cavendish Laboratory under Patrick Blackett and Ernest Rutherford to undertake a PhD in physics, studying the properties of the electron. At the height of the 1931 economic depression he took the only job he could find, as a demonstrator in the Department of Geodesy and Geophysics at Cambridge. Through his work in the department he became a consultant for Anglo-Iranian Oil, and in this capacity he began to adapt the secret seismic survey techniques used by oil companies for academic research. He realised that these surveying techniques would enable geologists to understand the sub-surface structure of the earth without digging boreholes and studying core samples of rock. In 1936, whilst studying the adaptability of industrial techniques, Bullard met US oceanographer Maurice Ewing, who was working on seabed composition using seismic techniques. This interaction introduced Bullard to oceanography, and prompted him to study and adapt these techniques for surveys alongside the Royal Navy, using geophones and hydrophones to create more accurate surveys of the seabed for Admiralty charts.[72]

Bullard and his colleagues in Cambridge made use of Royal Navy funds to conduct scientific work at sea before the Second World War. Since naval officers shared an interest in devices for surveys with the Cambridge scientists, the latter sought to use the instruments to gain new knowledge on the earth's structures, whereas the former realised their potential in minesweeping operations and charting. In 1938 the Cambridge Department for Geodesy and Geophysics convinced the Royal Navy to grant them the use of a submarine to conduct gravity surveys, using an instrument loaned by a Dutch professor, Dr Felix Andries Vening Meinesz.[73] These surveys were intended to be a part of an international effort to map the geomagnetic landscape of the seabed, a project that had begun in the early nineteenth century, with the object of conducting land-based magnetic surveys.[74] It had proved impossible to achieve the necessary stability to conduct these investigations at sea using a surface vessel, so a submarine was required.[75] The purpose of the surveys was to understand both geomagnetism and the geological composition of the seabed using seismic waves from underwater explosions to create seabed profiles. The use of submarines also allowed gravity measurements of submarine slopes and the taking of measurements as near to the sea floor as possible, as the ultimate aim was to investigate the geology below the earth's crust.[76] Vaughan's report *International Aspects of Oceanography* included a chart showing the extent of gravity surveys produced by a Dutch Navy submarine.[77]

The use of a submarine as a research vessel required the cooperation of scientists and naval authorities. To make the most of the loan of the gravity instruments, the head of the Cambridge department persuaded the Royal Navy to conduct two cruises, one led by Edward Bullard and one by Ben Browne (a fellow Cambridge geophysicist) and two submarines were made available with the assistance of the Hydrographic Office.[78] Bullard's voyage on HMS *Jason* began in July 1938 and undertook seismic work using explosives and geophones, which Cambridge had been developing during the previous summer at Windermere, to understand the geology of the sea floor. Browne's cruise aboard HMS

Narwhal, which began in September 1938, was more closely related to Vening Meinesz's work, but was cancelled two days into the voyage because of the Munich crisis.[79] This basic research on the magnetic properties of the earth was of little direct relevance to the war effort, but the skills, techniques, and knowledge of the Cambridge scientists were useful within the Mine Design Department (MDD) of the Royal Navy (Edward Bullard) and at HMS *Osprey* (Ben Browne). Here their "basic" knowledge of the geomagnetic properties of the earth could be applied to the construction of magnetic mines.

For the Admiralty the magnetic mine was the first major scientific challenge of the Second World War. The weapon was considered so threatening that Winston Churchill met Albert Beaumont Wood, Chief Scientist of the MDD, personally to be briefed on the potential implications of magnetic mines, and to inspect a recovered German mine for himself.[80] Wood had gained a first-class honours degree in physics from the University of Manchester in 1910, staying on to work as an assistant to Ernest Rutherford. Dissatisfied with academic life, he had taken the opportunity of the First World War to join the Admiralty's newly formed Board of Invention and Research with a recommendation from Rutherford, and he continued to pursue scientific research within the Admiralty between the wars.[81]

Whilst the magnetic mine caused panic amongst politicians, the MDD had been considering the problem for some time. In September 1939, the department began to work on demagnetising ships (degaussing) to render them incapable of triggering magnetic mines.[82] Magnetic mines continued to wash up on the British coastline during the winter and spring of 1939–40. Wood and his team became involved in defusing these devices, putting their own lives at risk in order to understand how the German mine worked.[83] The MDD was attempting to produce a more effective British version of the magnetic mine, one impervious to changes in the earth's magnetic field or enemy degaussing technologies, and Bullard became part of the degaussing department headed by physicist Stephen Butterworth.[84] This group was tasked with mapping magnetic variations around the globe using magnetometers and developing a system that accounted for the magnetic changes affecting ships crossing the equator north–south or south–north.[85]

After eighteen months at the MDD, Bullard became frustrated that, unlike Deacon's, his research was not being prioritised. Bullard felt that the Navy did not know how to handle scientists such as himself; he claimed this created the belief amongst men such as Albert Wood that they could not affect policy or get their ideas implemented.[86] In a clash of scientific research cultures—the academic on one side and the military on the other—Bullard issued a tirade against his new overseers. In a letter to Fowler, head of the advisory committee on Admiralty research, Bullard identified 'serious defects in the Admiralty Experimental Establishments particularly *Osprey*'. In this letter, which contained many crossings out and revised passages, Bullard described HMS *Osprey*'s Chief Scientist, B.S. Smith, as 'a first class obstructionist' who would 'not allow any ideas but his own to be heard and insists on work on quite hopeless schemes being carried on'.[87] It was not only the management of research that irked Bullard; he also described the cooperation from the naval side as 'quite inadequate'. The issue at hand was one of specialisation, but also of methods: both Cambridge and the MDD were attacking the same problems but separately, with little overlap in their approaches.

Whereas in a university such as Cambridge various research tasks were brought together under one department, each naval experimental facility was rigidly focused on military applications. Deacon and his *Osprey* colleagues had succeeded in convincing the Admiralty Research Committee that they should be tasked with acoustic research, whereas Bullard had failed to convince similar authorities that his team should be allowed to take their research in this direction. When he approached the Admiralty to carry out experiments to assist with the problem of proving that acoustic mines were viable, he was ignored. However once it was proved that acoustic mines did indeed exist and those in authority began to panic and look for solutions, *Osprey* was quick to begin research in an attempt to win the credit for developing the technology required to defend against the latest German threat.[88]

In his letter Bullard also singled out the 'ludicrous' notion of having a naval Captain 'in command'.[89] According to Bullard, the Experimental Captain at *Osprey*, William Poulett, never challenged his Chief Scientist, who had developed his career within the Navy and therefore strictly adhered to the Admiralty's research culture. Bullard implied that Poulett was a "yes man", but stopped short of stating that this was due to a lack of scientific understanding on the Captain's part. He was also quick to place the blame for all these deficiencies firmly at the feet of Charles Wright, the then Director of Scientific Research (DSR) at the Admiralty and the future head of the Royal Naval Scientific Service (RNSS). Wright was described as indifferent, letting 'things drift and can only be persuaded to take action when it appears that a worse row will occur if he does nothing'.[90] Bullard believed that there were civilian scientists who could perform much better than the institutionalised scientists of the pre-war era Admiralty research community, because civilian scientists were less hierarchical and were used to a greater degree of research freedom. In 1942 Smith, the Chief Scientist, was removed from HMS *Osprey,* and research seems to have accelerated following his departure, especially on issues which Smith had been particularly dismissive of, such as the need for short-range accurate tracking of enemy submarines.

Despite his fractious relationship with the Royal Navy hierarchy and the attitudes to scientific research conducted under its banner, in 1945 Bullard began to argue for a renewed research effort directed towards magnetic surveys. He argued that the Navy was in danger of losing the initiative in magnetic measurements at sea to the Royal Air Force.[91] In 1947 the Department of Geodesy and Geophysics at Cambridge under Bullard's direction undertook further magnetic surveys on board Royal Navy submarines following a review of Admiralty research into terrestrial magnetism and earth currents.[92] In this Bullard, unlike Deacon, set his wartime experience aside; he wished to resume his pre-war scientific studies alongside the Royal Navy, rather than continue his war work.

But while Bullard vented his frustration, Deacon managed to consolidate oceanographic studies in the Admiralty Research Laboratory, bringing together research scientists from the array of research establishments that conducted some form of oceanographic research within the Royal Navy. It was he, and not Bullard, who was able to show naval commanders the potential of investing more in military oceanography under the oceanographers' leadership. Deacon's temperament, and his experience of working mostly at sea in ships run along very similar lines to Royal Navy vessels meant that he had no difficult in adjusting to life in an hierarchical research environment. Furthermore, he knew how to manipulate such environments to his advantage, thanks to his service as Principal Scientific Officer in his last pre-war *Discovery II* expedition.

Oceanography at the Admiralty Research Laboratory (1944–45)

Upon his return from the United States in 1944 Deacon approached Albert Beaumont Wood to discuss expanding the work carried out by the oceanographers for the Navy and the possibility of forming a "wave group" or "oceanographic group" at the Admiralty Research Laboratory

(ARL).[93] The ARL had been formed after the First World War by bringing together all the groups that had worked on anti-submarine warfare around the UK during the conflict. Prior to 1945 the ARL was primarily focused on developing sonars and offensive anti-submarine weapons (depth charges), rather than the physical study of the ocean, which was based in the Hydrographer's office. On 5 June 1944 Deacon reported for duty at ARL as director of the swiftly formed Oceanography Group. The urgency with which this group was established derived from a new German threat, the "oyster mine" (which was triggered by the pressure wave of a ship passing overhead).[94] It was an advanced concept taken straight from land-based combat, where mines had been developed which required an object of a certain weight to trigger them, so they did not explode if a soldier stepped on them, but detonated if a heavy vehicle ran over them. Like many German weapons rushed into combat by late 1944, the oyster mine was not perfected, and the storms of spring 1944 detonated most of the mines that the Germans had deployed. The MDD focused on perfecting British weapons to avoid the issue of premature detonation by climatic conditions rather than by a passing ship.[95] Unlike with the degaussing of ships, there was no retrofit solution to the pressure mine; all moving ships created a pressure wave, and were therefore vulnerable.

The mine design issue could only be solved with a greater understanding of the pressure created naturally by waves and storm surges. In order to better understand these phenomena, Wood agreed that Deacon should establish wave-recording stations on the north and south coasts of Cornwall to measure these changes. As Wood put it in his memoirs, "our own design of 'oyster' was made to cope with probabilities in sea water".[96]

Deacon was now responsible for leading a small team (physicists Norman Barber, Jack Darbyshire, and M. J. 'Tom' Tucker, freshwater biologist Clifford Mortimer, and mathematician Fritz Ursell), which became the nucleus of physical oceanography within the Admiralty research community.[97] Amongst its ranks were various talented mathematicians, in addition to oceanographers and geophysicists. Fritz Ursell, who arrived as a German refugee in 1937 and studied mathematics at Cambridge on a shortened wartime course, was selected for research, and later became Professor of Applied Mathematics at the University of Manchester.[98]

Deacon's team reflected the composite nature of oceanography itself, bringing together specialists from physics, chemistry, engineering, mathematics, and pre-war fisheries science. This formula was later replicated at the National Institute of Oceanography, where scientists of various backgrounds were recruited before evolving into 'oceanographers'. Many of this team came from other groups at ARL, often from sections within group H,[99] which had worked on the degaussing programme, amongst others. By 1944 there were few scientists available for redirection to new programmes. The W (Wave) group was thus composed of predominantly young scientists who could be easily spared from other projects.[100] Ursell claimed, for instance, that he only became part of the W group because his old research group had had to move out of the room earmarked for the new team. Ursell's memory is probably slightly inaccurate; it appeared that Darbyshire and Barber had already suggested to Stephen Butterworth that Ursell might be more interested in moving with them to work on waves.[101] By 1945 the group had expanded as other personnel were released from their respective research sections.[102] The initial team expanded to include mathematician Michael Longuet-Higgins, a marine engineer, and two instrument makers. The diversity of skills within the group provided its scientific strength, enabling them to be almost self-sufficient as a research team able to create instruments, crunch data, and provide scientific interpretation all in-house. This was important as in their study of waves, both surface and internal, the group needed to create their own wave-measuring equipment, interpret their own data, and build this towards the creation of a theory of ocean waves that could feed into a British pressure-mine programme.

The W group was tasked with undertaking basic research for applied purposes. How internal waves affected pressure and sound propagation had been largely neglected by physical oceanographers. By 1945, in order to begin collecting data, wave-recording stations had been established at Pendeen and Perranporth in Cornwall. The instrument first deployed to measure waves was an American-made "powerphone", which recorded changes in pressure using a fluxmeter. In addition to this device, an aneroid pressure measurer was developed by the Cambridge Instrument Company and deployed at Perranporth.[103] However, these devices only worked in shallow water, and there was no instrument in existence that could measure waves in deep water. Deacon and his assistants foresaw that eventually by using buoys one could improve the remote sensing of oceanographic parameters, but this was not likely to happen before the end of the war. By March 1945 two different types of wave recorder were being developed at ARL for use in a buoy. One of these detectors was intended to record pressure in the same way as existing instruments, but when hanging from a line 100ft below the buoy. The other detector was used to record wave height based on the vertical accelerations of the buoy.[104]

Not only did the W group allow Deacon to set up his own team and strengthen ties with the Admiralty, it also put him in contact with intelligence personnel. In the case of oyster mines, the W group was asked whether there was any useful information they would like Naval Intelligence officers to extract from the German naval bases which had been captured at the end of the war.[105] The W group response reflected their wave research and the belief that before deploying the oyster mines, the Germans must have conducted some form of wave or pressure experiments.[106] In less than a month the W group received their reply from the intelligence services. The Directorate of Naval Intelligence (DNI) informed Deacon's team that wave research in Germany had been conducted using a pre-war wave buoy. Additionally, Naval Intelligence had rounded up all of the "personalities" involved at Travemünde; the DNI therefore requested that 'wave motion experts be sent from Admiralty to cover this'.[107] As shown at the beginning of this chapter, Carruthers did set out to interrogate his German colleagues after the war, eventually producing a report which was hard bound and handed over to the Admiralty.[108] The DNI supported Admiralty scientists in gaining knowledge of German research because it potentially had direct benefits for British defence. Additionally having British scientists who were knowledgeable about the subject make judgements on enemy developments made creating intelligence assessments much easier for the non-scientifically expert body.

The cessation of hostilities in Europe granted Deacon the opportunity to invite guests to come and visit the W group. In August 1945, one of the recognised celebrities in oceanography, Harald Sverdrup, Director of the Scripps Institute of Oceanography, one of the two leading oceanographic institutions in the USA, and Norman Jefferis Holter, a physicist who studied ocean wave characteristics for the US Navy's Bureau of Ships, visited to observe wave measurements themselves. This was also an opportunity for the British to showcase the advances they had made. In a letter to a naval captain who had assisted in the laying of equipment at Perranporth, Charles Seymour Wright, Director of Scientific Research at the Admiralty, boasted, 'A fortnight ago we were able to take the leading U.S. Oceanographer down to Perranporth and obtain the smug satisfaction of an admission that we had got our noses ahead of him in wave-measuring technique'.[109] This was corroborated in a letter sent to Deacon by Holter on his return to the United States. Holter informed Deacon that his work on waves was of great interest and requested further information, specifications, and prices for all the British equipment he had been shown in August.[110] Pointing towards the purchasing power of the US Navy, Holter also asked Deacon to supply a price for the outright purchase of the 'fancy model of the wheel analyser'.[111]

Holter's letter championed Woods Hole Oceanographic Institution (WHOI), the other leading US oceanographic institutes, as having the greatest interest in wave research, informing Deacon that they were attempting to construct an analyser themselves. This is confirmed in the later correspondence between the W group and the United States, which predominantly involved Holter and Columbus Iselin, Director of WHOI. In the USA the wave analyser eventually developed from a programme to build oceanographic instruments at the University of California at Berkeley.[112] Realising the significance of wave research for non-scientific audiences, especially serving naval officers, Iselin persuaded Henry Bigelow to publish a book on waves in 1947 entitled *Wind Waves at Sea, Breakers and Surf* which was translated into Russian in 1951.[113] This was a scientific intelligence-gathering achievement for the Soviets because US oceanographers had assured the US Navy that no classified or militarily useful American knowledge would be handed over so easily to the Soviets.[114] In the United Kingdom the W group instead produced a series of technical publications between 1945 and 1947, with extremely limited circulation due to their classified status.[115] Over time the British W group started to publish more generalised studies outside Admiralty research circles, most often in *Nature*, thus moving from a wartime military footing to more recognisably civilian practices.[116]

The circumstances that shaped the emergence of a military-scientific relationship between oceanographers, the Navy, and the Admiralty during the Second World War fundamentally changed how ocean science research was organised in Great Britain. Later in the Cold War, wartime experiences provided validation of the military expertise of civilian oceanographers, these wartime relationships and networks becoming an underlying feature of Cold War interactions and resulting in long-term collaborations between defence research and British oceanographers. The particular circumstances of the Second World War demanded scientific intelligence on enemy capabilities and intentions, scientific help to defeat submarines, and strategically relevant scientific knowledge for military operations. Importantly, the personalities and the expanding agency of individual actors were central to the construction of these relationships between producers of knowledge of the oceans and the users and adapters of that expertise.

Oceanographers were able to adapt their existing skills to military research. Following the war, many of the technologies and devices encountered during wartime service were deployed by oceanographers to further their own civilian research. Additionally, the facilities, resources, and finances of the Admiralty research establishments allowed oceanographers to experiment with new instruments and techniques. Deacon's directorship of the W group, Carruthers' position as the Hydrographer's unofficial assistant, and Bullard's senior position in the Operational Research Department enabled a great deal of influence for the oceanographers, allowing them into the Admiralty post-war policy discussions.

The war also brought leading American and British oceanographers into greater contact than had occurred previously. Cooperation earlier in the war on ASDIC generated American interest in Deacon's W group. Furthermore the war enabled rather than restricted travel, allowing many British oceanographers to make their first trips to the United States at the Admiralty's expense. Exposure to the American facilities impressed Deacon and others. Their desire to replicate the US oceanographic institutes of Woods Hole in Maine and Scripps in California animated debates surrounding the post-war planning for oceanography within the Royal Society during 1943–44, as discussed in Chap. 3. These interactions also facilitated the growth of relationships that ultimately led to collaboration on wave and other research during the Cold War.

George Deacon emerged as the one scientist most capable of building networks for the oceanographers; he excelled at mediating between the military and the community of civilian ocean scientists. James Carruthers demonstrated his ability to gather scientific intelligence, condensing this research into briefing papers that could be used by other scientists to develop solutions to strategic problems, but failed either to develop his own scientific programme or to emerge as a scientific administrator/manager. Edward Bullard, tasked with developing those solutions, could not find a sufficiently comfortable working atmosphere within the Admiralty research culture. Deacon found it much easier to circulate in scientific and military cultures; he did not need to adapt these cultures to his own style of working. In this way Deacon was a hybrid, able to unite his skills as a scientist with those of a negotiator between his research establishment and foreign colleagues, not as a scientific diplomat *per se*, but in a much more diplomatic and persuasive way than Bullard and Carruthers. He was the right "type" of man in the minds of senior Admiralty science administrators, who felt that he understood how science might be best applied to operational challenges. Hence he was the right person in the right place at the right time when post-war reorganisation of naval science was being discussed. This resulted in his elevation through the ranks of the Admiralty research division until he assumed a position of scientific leadership.

In analysing military-scientific synergies it becomes apparent that the Admiralty, itself influenced by the influx of civilian scientists into its ranks, eventually began to regard this relationship as mutually beneficial. Civilians would be able to do research for the Admiralty without the need for direction and similarly the Royal Navy would perhaps be able to supply the oceanographers with data collected from their measurements. The key characteristic of this relationship became its flexibility and ability to adapt to the requirements of the moment. Oceanographers had shown that their science had operational value and in return they had seen what the Navy's resources could potentially enable in terms of future research. However, as is shown in the next chapter, none of this was certain and wartime relationships had to be converted into peacetime partnerships if naval patronage was to continue to be at the centre of British oceanographic developments.

Notes

1. Quoted and translated in Ramster, J., "Dr. J.N. Carruthers 24 November 1895–8 March 1973," *Journal du Counseil Permanent International pour l'Exploration de la Mer,* 36 (1975): 103.
2. Sonar and echo-sounding systems were known in Britain as ASDIC. Although this is often referred to as an acronym with varying wording, its most likely that the term was coined by the Admiralty as nonsense intended to give away as little detail as possible about the device. The confusion probably stems from its first use in the House of Commons by Churchill in 1939. Hackmann, probably the most authoritative study on early ASW, states 'The Oxford University Press was prompted on 11 December 1939 to ask the Admiralty about its etymology after Churchill used the term in the House of Commons. After a certain amount of interdepartmental discussion, they were told that the word was the acronym of Allied Submarine Detection Investigation Committee, "a body which was formed during the war of 1914–1918, and which organised much research and experiment for the detection of submarines."' However, no such committee exists in the WW1 Admiralty records. See Willem

Hackmann, *Seek & Strike: Sonar, Anti-submarine Warfare and the Royal Navy 1914–54* (London: HMSO, 1984): xxv.
3. Margaret Deacon, *Scientists and the Sea 1650–1900: A Study of Marine Science* (Aldershot: Routledge, 1971).
4. "Re-challenge the Oceans," *Nature* (23 September 1920): 101–2.
5. Peder Roberts, *The European Antarctic: Science and Strategy in Scandinavia and the British Empire* (London: Palgrave Macmillan, 2011): 34–5; D. Graham Burnett, *The Sounding of the Whale: Science and Cetaceans in the Twentieth Century* (Chicago: University of Chicago Press, 2012): 107–111.
6. For Discovery Committee see Roberts, *The European Antarctic* 31–52; for transfer of vessels to NIO see Chapter Three.
7. For more detailed history of Lowestoft, see A. J. Southward, "History of the Marine Biological Association of the United Kingdom," *Helgoländer Meeresuntersuchungen,* 49, no. 1 (1995): 465–6.
8. See Chapter Two, Anna Carlson-Hyslop, "An Anatomy of Storm Surge Science at Liverpool Tidal Institute 1919–1959: Forecasting, Practices of Calculation and Patronage," (PhD diss., University of Manchester, 2010).
9. Thomas Wayland Vaughan, *International Aspects of Oceanography: Oceanographic Data and Provisions for Oceanographic Research* (Washington D.C.: National Academy of Sciences, 1937).
10. Vaughan was Director of the Scripps Institute of Oceanography until Harald Sverdrup replaced him in 1938.
11. Vaughan visited: England, Norway, Sweden, the Netherlands, France, Spain, Monaco, Germany, Denmark, Italy, Egypt, Siam (Thailand), French Indo-China (Vietnam), China, the Philippines, Japan, Hawaii, New Zealand, the East Indies, Australia, Malay (Malaysia).
12. Hamblin, *Oceanographers and the Cold War,* 30–1.
13. The challenges of cataloguing oceanographic data in the pre-digital age are outlined in Iouri Oliounie and Peter Pissierssens, "Oceanographic Data: from Paper to Pixels," in *Troubled Waters: Ocean Science and Governance,* ed. Geoff Holland and David Pugh (Cambridge: Cambridge University Press, 2010): 167–186.
14. Sabine Höhler, "Depth Records and Ocean Volumes: Ocean Profiling by Sound Technology, 1850–1930," *History and Technology,* 18 (2002): 119–154. Georg Wüst, "The Major Deep-Sea Expeditions and Research Vessels 1973–1960," *Progress in Oceanography,* 2 (1964):1–52.
15. Fellowship of the Royal Society was the highest honour for a British scientist. Fellows were elected by their peers, so a naval officer's election was rare and demonstrated the esteem in which Edgell was held within scientific circles.

16. Amongst many innovations Edgell was credited with replacing the lead line with the echo-sounder, and acquiring the funding to build the RRS *Research*, a wooden magnetic study vessel that was never launched due to the war. See George Deacon, "John Augustine Edgell. 1880–1962." *Biographical Memoirs of Fellows of the Royal Society*, 9 (1963): 87–90.
17. Margaret Deacon, "Steps Toward the Founding of NIO," in *Of Seas and Ships and Scientists: The Remarkable Story of the UK's National Institute of Oceanography*, ed. Anthony Laughton, John Gould, M.J. 'Tom' Tucker, Howard Roe (Cambridge: Lutterworth Press, 2010).
18. George E.R. Deacon, "Obituary 'James Norman Carruthers'", *Polar Record*, 16 (1973): 874.
19. Memo 'Projected British Oceanographical Institute', J.N. Carruthers, 6th October 1947, GERD Papers, M3/2/7, NOC Library (Southampton).
20. Deacon, "James Norman Carruthers," 874. Rear Admiral G. S. Ritchie, *No Day Too Long – An Hydrographer's Tale* (Edinburgh: The Pentland Press, 1992): 129.
21. On the concept of scientific middlemen, between science and engineering, see Takehiko Hashimoto, "Leonard Bairstow as a Scientific Middleman: Early Aerodynamic Research on Airplane Stability in Britain, 1909–1920," *Historia Scientiarum*, 17 (2007): 110 –112.
22. Memo 'Projected British Oceanographical Institute', J.N. Carruthers, 6 October 1947, GERD Papers, M3/2/7, NOC Library (Southampton).
23. Memo 'Projected British Oceanographical Institute', J.N. Carruthers, 6 October 1947, GERD Papers, M3/2/7, NOC Library (Southampton). NOC Library (Southampton).
24. Ibid.
25. By the end of the war Carruthers had been appointed Superintendent of the Oceanographic Branch of the Hydrographic Office, a branch created for him by the Navy. This continued after Carruthers became Assistant Director of the NIO under the leadership of a naval officer, Captain Ritchie, later Royal Navy Hydrographer. See Ritchie, *No Day Too Long*, 128–131.
26. Deacon, "James Norman Carruthers," 874.
27. Memo 'Projected British Oceanographical Institute', J.N. Carruthers, 6 October 1947, GERD Papers, M3/2/7, NOC Library (Southampton).
28. Ibid.
29. Ibid.
30. A complete list of scientists working in the Hydrographic Office has been difficult to obtain, but Carruthers was not the only expert recruited into the department. Geographers were also in high demand; see Avril Maddrell, "The 'Map Girls'. British women geographers' war work, shifting

gender boundaries and reflections on the history of geography," *Transactions of the Institute of British Geographers,* 33 (2008): 127–148.
31. See discussion of the cost savings of networks in Chap. 1, p. 9.
32. Boarding Card HMS *Kingfisher* [underlining in original], January 1940, GERD Papers, C1/1, NOC Library (Southampton).
33. Dominic Berry, "Genetics, Statistics, and Regulation at the National Institute of Agricultural Botany, 1919–1969," (PhD diss., University of Leeds, 2014), Chapter four, 'National Institute: Agricultural science in the Second World War, 1939–1955'. Provides an overview of British food supply problems during the conflict.
34. M. Fortun and S. S. Schweber, "Scientists and the Legacy of Second World War: The Case of Operations Research (OR)," *Social Studies of Science* 23 (1993): 595–642; Mary J. Nye, *Blackett: Physics, War, and Politics in the Twentieth Century* (Cambridge, MA: Harvard University Press, 2004): 75–85; Malcolm Llewellyn-Jones, "A Clash of Cultures: The Case for Large Convoys," in Peter Hore, *Patrick Blackett: Sailor, Scientist, Socialist* (London: Routledge, 2004): 137–166.
35. Lawrence Sondhaus, *The Great War at Sea: A Naval History of the First World War* (Cambridge: Cambridge University Press, 2014): 241–277.
36. Stephen Broadberry and Peter Howlett, "The United Kingdom: 'Victory at all costs'," in *The Economics of World War II: Six Great Powers in International Comparison,* ed. Mark Harrison (Cambridge: Cambridge University Press, 1998).
37. Malcolm Llewellyn-Jones, *The Royal Navy and Anti-Submarine Warfare, 1917–49* (London: Routledge, 2006); Donald P. Steury, "Naval Intelligence, the Atlantic Campaign and the Sinking of the *Bismarck:* A Study in the Integration of Intelligence into the Conduct of Naval Warfare," *Journal of Contemporary History,* 22 (1987): 209–233; Christopher M. Bell, *The Royal Navy, Seapower and Strategy between the Wars* (Redwood City, CA: Stanford University Press, 2000).
38. Williem Dirk Hackmann, *Seek & Strike: Sonar, Anti-Submarine Warfare and the Royal Navy 1914–54* (London: HMSO, 1984): 97–158; Williem Dirk Hackmann, "Underwater acoustics and the Royal Navy, 1893–1930," *Annals of Science* 36 (1979): 255–278; Williem Dirk Hackmann, "Sonar research and naval warfare 1914–1954: A case study of a twentieth-century establishment science," *Historical Studies in the Physical and Biological Sciences,* 16 (1986): 83–110.
39. Known in Britain as ASDIC.
40. Julie K. Peterson, *Understanding Surveillance Technologies: Spy Devices, Privacy, History & Applications* (New York: Auerbach Publications, 2007): 231.

41. Discussed below; for early history of ASDIC, see Hackmann, *Seek & Strike*, 73–96.
42. Hamblin, *Oceanographers and the Cold War*, 40–1.
43. Description of Fairlie Underwater Range, February 1942, ADM 259/571, TNA (London).
44. After the war this group established a sonar research centre in Norway as part of the Forsvarets Forskningsinstitutt (the Defence Research Institute); see Njølstad, O., Wicken, O., *Kunnskap som våpen: forsvarets forskningsinstitutt 1946–1975* (Oslo: Tano Aschehoug, 1997): 36–37. Anti-submarine warfare: staff targets for research and developments, ADM 1/13711, TNA (London).
45. Simone Turchetti, "Sword, Shield and Buoys: A History of the NATO Sub-Committee on Oceanographic Research 1959–1973," *Centaurus*, 54 (2012): 205–231.
46. See ADM 1/13711, and Hackmann, *Seek & Strike*, 267–302.
47. George R. Ehrhardt, "Bathythermograph," in *Instruments of Science: An Historical Encyclopedia*, ed. Robert Bud and Deborah Warner (London: Garland Publishing, 1998): 54–56.
48. Gary E. Weir, *An Ocean in Common; American Naval Officers, Scientists, and the Ocean Environment* (College Station: Texas A&M UP, 2001): 163.
49. Henry Wood had worked before the war at the Marine Laboratory of the Fishery Board for Scotland (Aberdeen), and was listed as HMS *Osprey*'s Senior Naturalist in the Navy List.
50. "Bathythermograph and ASDIC Observations on Convoy route to North Russia. HMS "Savage" October-November 1943", Internal Report 169, H. Wood, ADM 259/614, TNA (London).
51. Ibid.
52. Ibid.
53. Ibid., Report dated 10 November, Wood returned on 11 November.
54. "Preliminary Report on the Effect of Temperature Gradients on the Accuracy of Depth Measurements by ASDIC apparatus," 10 November 1943, G. Deacon and H. Wood, GERD Papers C1/5, NOC Library (Southampton).
55. James R. Fleming, "Sverre Petterseen, the Bergen School, and the Forecasts for D-Day," *Proceedings of the International Commission on History of Meteorology* 1 (2004): 75–83.
56. "Bathythermograph and ASDIC Observations on Convoy route to North Russia. HMS "Savage" October-November 1943", Internal Report 169, H. Wood, ADM 259/614, TNA (London).
57. Minute attached to Internal Report 169 by Captain Third Destroyer Flotilla: 'data made available by bathythermograph observations "on the

spot" will be of tremendous value in A/S operations and in this connection assistance in taking observations and keeping records will be afforded whenever practicable,' ADM 259/614, TNA (London).

58. Richard Hammond, "British Policy on Total Maritime Warfare and the Anti-shipping Campaign in the Mediterranean, 1940–1944," *Journal of Strategic Studies* 36:6 (2013): 789–814; Barbara Brooks Tomblin, "The Naval War in the Mediterranean," in Thomas W. Zeiler, Daniel M. DuBois, *A Companion to World War II* (Edinburgh: Blackwell Publishing, 2013): 222–242.
59. Dick, R.M., "The Navy's Part in the North African Campaign," *Royal United Services Institution Journal,* 89:555 (1944): 261–274; Atilano, J.A., "The Transatlantic Essay Contest and the Planning Principles of the North African Campaign," dtic online. URL: http://oai.dtic.mil/oai/oai?verb=getRecord&metadataPrefix=html&identifier=ADA583732 (accessed 9/12/14).
60. "First Mediterranean Patrol Report for period 13th–30th June 1943", Commanding Officer HMS 'Templar', ADM 1/15410, TNA (London).
61. Weir, *An Ocean in Common,* 194.
62. Note, Commander-in-Chief Mediterranean, 13 August 1943, ADM 1/15410, TNA (London).
63. Minute from Admiral (Submarines), 4 September 1943, ADM 1/15410, TNA (London).
64. Ibid.
65. Ibid.
66. Draft of "Use of Bathythermograph in British Submarines," GERD Papers, C1/6, NOC Library (Southampton).
67. Comments by USA on GERD authored, ASW Fairlie report 127 (10 May 1943), GERD Papers, C1/6, NOC Library (Southampton).
68. Hackmann, *Seek & Strike,* 261–3.
69. Weir, *An Ocean in Common,* 164–5.
70. Report 174: "Recent Developments in the Study of the effect of water conditions on ASDIC performance and their applications to submarine and anti-submarine Warfare," December 1943, GERD Papers, C1/8, NOC Library (Southampton).
71. "The Battle of the Atlantic and Signals Intelligence: U-boat Situations and Trends, 1941–1945," *Publications of the Navy Records Society Vol.139* (Farnham, UK: Ashgate, 1998): 186. The post-war official Royal Navy history of the Battle of the Atlantic stated that in the eyes of the Royal Navy May 1943 was the turning point in the battle against merchant shipping losses in the North Atlantic; Naval Staff History (Second World War) "The defeat of the enemy attack on shipping 1939–1945; A study of Policy and operations." (Admiralty, 1957): 228.

72. D.P. McKenzie, "Edward Crisp Bullard. 21 September 1907–3 April 1980," *Biogr. Mems Fell. R. Soc.,* (1987) 66–98; R.J. Howarth, "Bullard, Sir Edward Crisp (1907–1980)," *Oxford Dictionary of National Biography*. URL: <http://www.oxforddnb.com/view/printable/30870> (accessed: 9/12/14).
73. Discussion at the Admiralty, 21 July 1938, ADM 1/19723, TNA (London).
74. John Carwood, "The Magnetic Crusade: Science and Politics in Early Modern Britain," *Isis,* 70:4 (1979): 492–518.
75. Discussion at the Admiralty, 21 July 1938, ADM 1/19723, TNA (London).
76. Vening Meinesz, F.A., "Results of Maritime Gravity Research," in Vaughan, *International Aspects,* 61–66.
77. Vaughan, *International Aspects,* 59–69.
78. Cambridge Annual Report 1939, ADM 1/19723, TNA (London).
79. Ibid.
80. Albert Beaumont Wood, "From the Board of Invention and Research to the Royal Naval Scientific Service," *Journal of the Royal Naval Scientific Service,* 20:4 (July, 1965): 262.
81. W. Laurence Borrows, "Albert Beaumont Wood, O.B.E., DSc, Memorial Number," *Journal of the Royal Naval Scientific Service,* 20:4 (July, 1965): 188. On the Admiralty Board of Invention and Research, see Roy M. MacLeod and Kay E. Andrews, "Scientific Advice in the War at Sea, 1915–1917: The Board of Invention and Research," *Journal of Contemporary History* 6, no. 2 (1971): 3–40.
82. Albert Beaumont Wood, "Growth of Civilian Scientific Research in the Royal Navy, 1937–45," ADM 218/7, TNA (London).
83. Wood, "From the Board of Invention and Research to the Royal Naval Scientific Service," 263–4.
84. McKenzie, "Edward Crisp Bullard," 75–6.
85. Albert Beaumont Wood, "Growth of Civilian Scientific Research in the Royal Navy, 1937–45," ADM 218/7, TNA (London).
86. Edward C. Bullard, "Albert Beaumont Wood, O.B.E., DSc, Memorial Number," *Journal of the Royal Naval Scientific Service,* 20:4 (July, 1965): 191.
87. Letter from E.C. Bullard to Professor Fowler, 9 June 1941, E3, CSAC 100.4.84, CAC. This episode is also highlighted in David Edgerton, *Britain's War Machine: Weapons, Resources and Experts in the Second World War* (Oxford: Oxford University Press, 2011) 247.
88. Letter from E.C. Bullard to Professor Fowler, 9 June 1941, E3, CSAC 100.4.84, CAC.

89. This was a regular bone of contention in the Royal Navy during the war; see C.I. Hamilton, "Three cultures at the Admiralty, c.1800–1945: Naval Staff, the Secretariat, and the arrival of scientists," *Journal for Maritime Research,* 16:1 (2014) 98.
90. Ibid.
91. Bullard to the Director of Naval Operational Research, 14 May 1945, ADM 1/19723, TNA (London).
92. "Recent Experiments by the Admiralty in Terrestrial Magnetism and Earth Currents," 6 August 1948, ADM 204/861, TNA (London).
93. Albert Beaumont Wood, "Growth of Civilian Scientific Research in the Royal Navy, 1937–45," ADM 218/7, TNA (London).
94. Howard S. Levie, *Mine Warfare at Sea* (Amsterdam: Martinus Nijhoff Publishers, 1992): 110.
95. Dr. A. B. Wood, "The War Period Part V; From B.I.R. to R.N.S.S. 1937–1945," ADM 218/7, TNA (London).
96. Ibid.
97. Anthony Laughton and Margaret Deacon, "The founding director: Sir George Deacon," in *Of Seas and Ships and Scientists: The Remarkable Story of the UK's National Institute of Oceanography,* ed. Anthony Laughton, John Gould, M.J. 'Tom' Tucker, Howard Roe (Cambridge: Lutterworth Press, 2010): 37.
98. Michael Longuet -Higgins, "Group W at the Admiralty Reseach Laboratory," in *Of Seas and Ships and Scientists: The Remarkable Story of the UK's National Institute of Oceanography,* ed. Anthony Laughton, John Gould, M.J. 'Tom' Tucker, Howard Roe (Cambridge: Lutterworth Press, 2010).
99. H for Hydrodynamics.
100. Fritz Ursell, section edited by Michael Longuet-Higgins, "Group W at the Admiralty Research Laboratory," in *Of Seas and Ships and Scientists: The Remarkable Story of the UK's National Institute of Oceanography,* ed. Anthony Laughton, John Gould, M.J. 'Tom' Tucker, Howard Roe (Cambridge: Lutterworth Press, 2010): 42.
101. Jack Darbyshire, section edited by Michael Longuet-Higgins, "Group W at the Admiralty Research Laboratory," in *Of Seas and Ships and Scientists: The Remarkable Story of the UK's National Institute of Oceanography,* ed. Anthony Laughton, John Gould, M.J. 'Tom' Tucker, Howard Roe (Cambridge: Lutterworth Press, 2010): 48.
102. Ibid.
103. "Oceanographical Group ARL" Memo 21 March 1945, GERD Papers, C6/2, NOC Library (Southampton).
104. "Oceanographical Group ARL" Memo 21 March 1945, GERD Papers, C6/2, NOC Library (Southampton).

105. British Naval Intelligence was one of the most important of the British intelligence-gathering services during the war, and was particularly focused on gathering scientific and technical information. See Roger Victor Jones, *Most Secret War* (London: Penguin, 2009); James M. Goodchild, "R.V. Jones and the Birth of Scientific Intelligence," (PhD diss., University of Exeter, March 2013).
106. Letter N.Barber, 4 June 1945, GERD Papers, C6/2, NOC Library (Southampton).
107. Letter from Lt. Cameron R.N.V.R to Admiralty D.N.I. for ARL Wave Group, 24th June 1945, GERD Papers, C6/2, NOC Library (Southampton).
108. A copy was also published in the *Transactions: American Geophysical Union*; see Boehnecke, G., "German Oceanographic Work, 1934–1945," *Transactions: American Geophysical Union*, 29:1 (1948): 59–68.
109. Letter from Wright to Cpt. Cooper, 21 September 1945, GERD Papers, C6/2, NOC Library (Southampton).
110. Letter N.J. Holter, US Navy Bureau, GERD Papers, C6/2, NOC Library (Southampton).
111. Ibid., GERD Papers, C6/2, NOC Library (Southampton).
112. Memorandum Columbus Iselin to Dr. A.N. Richards and Dr. Detlev W. Bronk,1 September 1948, Columbus O'Donnell Iselin papers, 1904–1971. MC-16, "Memo to Dr. A.N. Richards and Dr. Detlev Bronk." Data Library and Archives, Woods Hole Oceanographic Institution.
113. Henry B. Bigelow and W.T. Edmondson, *Winds Waves at Sea, Breakers and Surf*, (Washington D.C.: US Govt. Print, 1947). [DOI: https://doi.org/10.5962/bhl.title.34810].
114. There were strong links between Soviet and American oceanographers at the time. As shown by Peder Roberts, the President of the American National Academy of Science had a longstanding relationship with Alexansder Nikolaevich Nesmeyanov, President of the AoS (Academy of Sciences of the USSR), who together established a framework for exchange despite the sensitive nature of the scientific work at the geo-physics–military interface; see Peder Roberts, "Scientists and Sea Ice under Surveillance in the Early Cold War," in *Surveillance Imperative*, ed. Turchetti, Roberts, 125–146.
115. "Wave Characteristics at Perranporth, 1946," ADM 204/2097, TNA (London); "Generation of Ocean Waves and Swell, 1947," ADM 213/575, TNA (London); "The effect of the rotation of the earth on the propagation of waves, 1947," ADM 204/2106, TNA (London).
116. Norman Barber and Fritz Ursell, "Study of Ocean Swell," *Nature* 159: 4032 (1947).

References

Barber, Norman, Fritz Ursell. "Study of Ocean Swell." *Nature*. 8 February 1947.

Bell, Christopher M. *The Royal Navy, Seapower and Strategy Between the Wars.* Redwood City, CA: Stanford University Press, 2000.

Berry, Dominic. "Genetics, Statistics, and Regulation at the National Institute of Agricultural Botany, 1919–1969." PhD diss., University of Leeds, 2014.

Bigelow, Henry B., W.T. Edmondson. *Winds Waves at Sea, Breakers and Surf.* Washington D.C.: US Govt. Print, 1947.

Boehnecke, Gunter. "German Oceanographic Work, 1934–1945." *Transactions: American Geophysical Union* 29 (1948): 59–68.

Borrows, W. Laurence. "Albert Beaumont Wood, O.B.E., DSc, Memorial Number." *Journal of the Royal Naval Scientific Service* 20 (1965): 188.

Broadberry, Stephen, Peter Howlett. "The United Kingdom: 'Victory at All Costs'." In *The Economics of World War II: Six Great Powers in International Comparison.* Edited by Mark Harrison. Cambridge: Cambridge University Press, 1998.

Bullard, Edward C. "Albert Beaumont Wood, O.B.E., DSc, Memorial Number." *Journal of the Royal Naval Scientific Service* 20 (1965).

Burnett, D. Graham. *The Sounding of the Whale: Science and Cetaceans in the Twentieth Century.* Chicago: University of Chicago Press, 2012.

Carlson-Hyslop, Anna. "An Anatomy of Storm Surge Science at Liverpool Tidal Institute 1919–1959: Forecasting, Practices of Calculation and Patronage." PhD diss., University of Manchester, 2010.

Carwood, John. "The Magnetic Crusade: Science and Politics in Early Modern Britain." *Isis* 70 (1979): 492–518.

Darbyshire, Jack. "Group W at the Admiralty Research Laboratory." In *Of Seas and Ships and Scientists: The Remarkable Story of the UK's National Institute of Oceanography.* Edited by Anthony Laughton, John Gould, M.J. 'Tom' Tucker, Howard Roe. Cambridge: Lutterworth Press, 2010.

Deacon, George E.R. "John Augustine Edgell. 1880–1962." *Biographical Memoirs of Fellows of the Royal Society* 9 (1963): 87–90.

Deacon, George E.R. "Obituary 'James Norman Carruthers'." *Polar Record* 16 (1973): 873–875.

Deacon, Margaret. *Scientists and the Sea 1650–1900: A Study of Marine Science.* Aldershot: Routledge, 1971.

Deacon, Margaret. "Steps Toward the Founding of NIO." In *Of Seas and Ships and Scientists: The Remarkable Story of the UK's National Institute of Oceanography.* Edited by Anthony Laughton, John Gould, M.J. 'Tom' Tucker, Howard Roe. Cambridge: Lutterworth Press, 2010.

Dick, R.M. "The Navy's Part in the North African Campaign." *Royal United Services Institution Journal* 89 (1944): 261–274.

Edgerton, David. *Britain's War Machine: Weapons, Resources and Experts in the Second World War.* Oxford: Oxford University Press, 2011.
Ehrhardt, George R. "Bathythermograph." In *Instruments of Science: An Historical Encyclopaedia*. Edited by Robert Bud, Deborah Warner. London: Garland Publishing, 1998.
Fleming, James R. "Sverre Petterseen, the Bergen School, and the Forecasts for D-Day." *Proceedings of the International Commission on History of Meteorology* 1 (2004): 75–83.
Fortun, M., S. S. Schweber. "Scientists and the Legacy of Second World War: The Case of Operations Research (OR)." *Social Studies of Science* 23 (1993): 595–642.
Goodchild, James M. "R.V. Jones and the Birth of Scientific Intelligence." PhD diss., University of Exeter, 2013. British Library EthOS (uk.bl.ethos.579986).
Hackmann, Williem Dirk. "Underwater Acoustics and the Royal Navy, 1893–1930." *Annals of Science* 36 (1979): 255–278.
Hackmann, Williem Dirk. *Seek & Strike: Sonar, Anti-Submarine Warfare and the Royal Navy 1914–54.* London: HMSO, 1984.
Hackmann, Williem Dirk. "Sonar Research and Naval Warfare 1914–1954: A Case Study of a Twentieth-Century Establishment Science." *Historical Studies in the Physical and Biological Sciences* 16 (1986): 83–110.
Hamblin, Jacob D. *Oceanographers and the Cold War: Disciples of Marine Science.* Seattle: University of Washington, 2005.
Hamilton, C.I. "Three Cultures at the Admiralty, c.1800–1945: Naval Staff, the Secretariat, and the Arrival of Scientists." *Journal for Maritime Research* 16, no.1 (2014):89–102.
Hammond, Richard. "British Policy on Total Maritime Warfare and the Anti-Shipping Campaign in the Mediterranean, 1940–1944." *Journal of Strategic Studies* 36 (2013): 789–814.
Hashimoto, Takehiko. "Leonard Bairstow as a Scientific Middleman: Early Aerodynamic Research on Airplane Stability in Britain, 1909–1920." *Historia Scientiarum* 17 (2007): 103 –120.
Höhler, Sabine. "Depth Records and Ocean Volumes: Ocean Profiling by Sound Technology, 1850–1930." *History and Technology* 18 (2002): 119–154.
Howarth, R.J. "Bullard, Sir Edward Crisp (1907–1980)." *Oxford Dictionary of National Biography.* Url: http://www.oxforddnb.com/view/printable/30870 (accessed: 9/12/14).
Jones, Roger Victor. *Most Secret War.* London: Penguin, 1977.
Laughton, Anthony, Margaret Deacon. "The Founding Director: Sir George Deacon." In *Of Seas and Ships and Scientists: The Remarkable Story of the UK's National Institute of Oceanography.* Edited by Anthony Laughton, John Gould, M.J. 'Tom' Tucker, Howard Roe. Cambridge: Lutterworth Press, 2010.

Levie, Howard S. *Mine Warfare at Sea*. Amsterdam: Martinus Nijhoff Publishers, 1992.
Llewellyn-Jones, Malcolm. "A Clash of Cultures: The Case for Large Convoys." In *Patrick Blackett: Sailor, Scientist, Socialist*. Edited by Peter Hore. London: Routledge, 2004.
Llewellyn-Jones, Malcolm. *The Royal Navy and Anti-Submarine Warfare, 1917–49*. London: Routledge, 2006.
Longuet-Higgins, Michael. "Group W at the Admiralty Reseach Laboratory." In *Of Seas and Ships and Scientists: The Remarkable Story of the UK's National Institute of Oceanography*. Edited by Anthony Laughton, John Gould, M.J. 'Tom' Tucker, Howard Roe. Cambridge: Lutterworth Press, 2010.
MacLeod, Roy M., Kay E. Andrews. "Scientific Advice in the War at Sea, 1915–1917: The Board of Invention and Research." *Journal of Contemporary History* 6 (1971): 3–40.
Maddrell, Avril. "The 'Map Girls'. British Women Geographers' War Work, Shifting Gender Boundaries and Reflections on the History of Geography." *Transactions of the Institute of British Geographers* 33 (2008): 127–148.
McKenzie, D.P. "Edward Crisp Bullard. 21 September 1907–3 April 1980." *Biographical Memoirs of Fellows of the Royal Society* 33 (1987): 66–98.
Nature. "Re-challenge the Oceans." 30 September 1920.
Njølstad, O., Wicken, O. *Kunnskap som våpen: forsvarets forskningsinstitutt 1946–1975*. Oslo: Tano Aschehoug, 1997.
Nye, Mary J. *Blackett: Physics, War, and Politics in the Twentieth Century*. Cambridge, MA: Harvard University Press, 2004.
Oliounie, Iouri, Peter Pissierssens. "Oceanographic Data: From Paper to Pixels." In *Troubled Waters: Ocean Science and Governance*. Edited by Geoff Holland and David Pugh. Cambridge: Cambridge University Press, 2010.
Peterson, Julie K. *Understanding Surveillance Technologies: Spy Devices, Privacy, History & Applications*. New York: Auerbach Publications, 2007.
Ramster, J. "Dr. J.N. Carruthers 24 November 1895–8 March 1973." *Journal du Conseil Permanent International pour l'Exploration de la Mer* 36 (1975): 101–105.
Ritchie, Real Admiral, G. S. *No Day Too Long – An Hydrographer's Tale*. Edinburgh: The Pentland Press, 1992.
Roberts, Peder. *The European Antarctic: Science and Strategy in Scandinavia and the British Empire*. London: Palgrave Macmillan, 2011.
Roberts, Peder. "Scientists and Sea Ice Under Surveillance in the Early Cold War." In *The Surveillance Imperative: Geosciences During the Cold War and Beyond*. Edited by Simone Turchetti, Peder Roberts. London: Palgrave Macmillan, 2014.
Sondhaus, Lawrence. *The Great War at Sea: A Naval History of the First World War*. Cambridge: Cambridge University Press, 2014.

Southward, A. J. "History of the Marine Biological Association of the United Kingdom." *Helgoländer Meeresuntersuchungen* 49 (1995): 465–6.

Steury, Donald P. "Naval Intelligence, the Atlantic Campaign and the Sinking of the *Bismarck:* A Study in the Integration of Intelligence into the Conduct of Naval Warfare." *Journal of Contemporary History* 22 (1987): 209–233.

Tomblin, Barbara Brooks. "The Naval War in the Mediterranean." In *A Companion to World War II.* Edited by Thomas W. Zeiler, Daniel M. DuBois. Edinburgh: Blackwell Publishing, 2013.

Turchetti, Simone. "Sword, Shield and Buoys: A History of the NATO Sub-Committee on Oceanographic Research 1959–1973." *Centaurus* 54 (2012): 205–231.

Ursell, Fritz. "Group W at the Admiralty Research Laboratory." In *Of Seas and Ships and Scientists: The Remarkable Story of the UK's National Institute of Oceanography.* Edited by Anthony Laughton, John Gould, M.J. 'Tom' Tucker, Howard Roe. Cambridge: Lutterworth Press, 2010.

Vaughan, Thomas Wayland. *International Aspects of Oceanography: Oceanographic Data and Provisions for Oceanographic Research.* Washington, D.C.: National Academy of Sciences, 1937.

Vening Meinesz, F.A. "Results of Maritime Gravity Research." In *International Aspects of Oceanography: Oceanographic Data and Provisions for Oceanographic Research.* Edited by Thomas Wayland Vaughan. Washington, D.C.: National Academy of Sciences, 1937.

Weir, Gary E. *An Ocean in Common; American Naval Officers, Scientists, and the Ocean Environment.* College Station, TX: Texas A&M University Press, 2001.

Wood, Albert Beaumont. "From the Board of Invention and Research to the Royal Naval Scientific Service." *Journal of the Royal Naval Scientific Service* 20 (1965): 185–284.

Wüst, Georg. "The Major Deep-Sea Expeditions and Research Vessels 1973–1960." *Progress in Oceanography* 2 (1964): 1–52.

CHAPTER 3

De-mobbing Military Oceanography: Post-War Needs of British Science

The Second World War was both an interruption and an opportunity for ocean science. The war saw a rapid expansion in the resources available to oceanographers who also gained first-hand experience of the Royal Navy research establishments. However, the war also demonstrated the limitations of oceanographic knowledge and the deficiencies in the pre-war organisation of oceanography. For universities the war caused an interruption in traditional training and research programmes, as departments began offering short fast-track courses for naval officers. Meanwhile their own staff went off into the military research establishments or into combat units. So by the end of the conflict it was unclear what their research and teaching focus would be.

The "de-mobbing" process produced debates around the key issues of funding, governance, and the research programmes for ocean science. For example, the future of the Discovery Committee remained unresolved: should it stay under the control of the Colonial Office and continue its imperialistic function, or should it be absorbed into the Admiralty, where its presence in the Southern Ocean could augment a naval presence in the region? Conversely, plans for the establishment of a National Institute of Oceanography (NIO) that would take responsibility for sea studies broadly construed, including military-related research, were now considered for the first time.

S. A. Robinson, *Ocean Science and the British Cold War State*,
Palgrave Studies in the History of Science and Technology,
https://doi.org/10.1007/978-3-319-73096-7_3

These issues had primarily to do with those officials in the British government who were responsible for understanding the requirements of a scientific discipline, weighed against the wider needs and capabilities of a country recovering from a long war. The Labour government of Clement Attlee, which was elected in July 1945, did not attempt to centralise scientific policy making. Nor did it try to bring about consensus when there was little to be had between scientists.[1] Analysis of the establishment of the NIO shows that the government was, rather, divided on these issues. Debates about the future of oceanography took place within the corridors of power, not in laboratories or on board ships. Only a small group of oceanographers were able to gain access to these corridors. Although their voices were restricted, they used them to good effect within the Admiralty, lobbying officers amongst the scientific staff and offering swift modification or clarifications to the potential research programme when required. The debates that occupied the oceanographers among themselves were at more of a personal and institutional level: Who should be director? Where should the Institute be located? Who should lead the first NIO expedition? What aspect of oceanographic research merits the primary attention of the Institute? These debates were secondary compared with the fundamental questions of whether the government or the military were prepared to fund new investigations. However, central to these funding decisions was the perceived balance of biological and physical studies within the disciplinary make-up of the new oceanography institution. How this was actively negotiated, and the decisions taken in this period, caused division within the ocean science community for many years but was also key to persuading funders and maintaining broad scientific support.

As part of these discussions, Deacon could now capitalise on a reputation gained during the war by entering the corridors of power and campaigning for investment in oceanographic research. In doing this he ended up representing government positions amenable to the Admiralty. Consequently as a "node" in the embryonic military oceanography network, he connected ocean science to government and, through these new links, re-configured existing ties between government and naval officers. It was not just because of Deacon's skills or personal qualities that he persuaded government officials, but rather because he represented new sets of interests that would allow a new policy network to expand and thrive. Understanding Deacon's biography and the networks within which he participated enables us to understand how the British government arrived at key decisions in relation to military oceanography.

The chapter is structured into three "rounds", a term used by civil servants in the Admiralty during 1947 when preparing briefing notes on the establishment of the NIO.[2] The first round focuses on the initial proposals for a post-war NIO, describing the circumstances surrounding these proposals and the responses of the various government departments and committees involved. It shows how during this early phase the proposals—which had seen such positive support from across Whitehall—became de-railed when a physical commitment of resources was required from the relevant departments.

The second round focuses on the years 1947–48, a period when the scientists themselves had to step in to persuade the Admiralty that any institution would place at its heart military oceanography, rather than surveying work. During this period the deconstruction of the earlier proposals, debates over financial responsibility for the proposed NIO, and the issue of funding basic research, almost led to the complete scrapping of the scheme.

The third round considers the steps taken after the passing of the Royal Charter that ratified the establishment of the NIO in April 1949. It is argued that this involved the re-building of the scheme put forward during the first round, which had been slowly unpicked by the debates and actions of various forces within Whitehall in 1946–48. This process of reassembly resulted in a scheme that closely resembled the original scheme of 1944. The role of the National Oceanographic Council is also considered here, as the body was designed to provide much needed governance for the NIO. This body was crucial in maintaining the civilian nature of the National Institute and for making sure that while completing tasks for the Admiralty it would continue to operate within the realm of fundamental research.

"First Round" (1944–46): Planning for the Post-war Needs of Science

At the end of the Second World War, Sir John Edgell, who as we have seen had played a key role in bringing oceanographers into wartime naval research, realised that the government had yet to design a policy catering for the continuation and development of military oceanography after the conflict.

The opportunity to instigate a debate on a future policy came through the Royal Society, of which Edgell became a Fellow in 1943. The Society had a historical role in promoting and encouraging the development and expansion of science in Britain, whilst remaining independent from government. By the latter stages of the war, however, the Royal Society had become entwined with the highest echelons of the British "war machine" and had launched a major study of the post-war needs of science in Britain.[3]

Edgell was commissioned to write a report in 1944 outlining the needs of post-war ocean science in Britain, after having been elected chair of the oceanography sub-committee of the Royal Society's Committee for Geophysics and Geodesy. He had been the Royal Naval Hydrographer since 1932, and as a result of his wartime experiences now considered oceanography to be fundamental to modern naval conflict. During the 1930s he had strengthened and rebuilt the surveying service, and was the first Hydrographer to be made a Fellow of the Royal Society in over a hundred years.

Edgell composed two reports simultaneously, one for the Admiralty and one for the Royal Society. Both had the same purpose: to find sponsors for sea studies.[4] Edgell wanted to further wartime oceanographic work without the entire costs of this continuation being assigned to the Admiralty. His reports thus aimed at "passing the parcel" of funding military oceanography to others in government—problems included submarine detection and beach profiling for amphibious landings. Ultimately, Edgell argued, there was a lack of knowledge of the modern hydrography of areas beyond Britain's traditional spheres of influence, and a general lack of understanding of the dynamics of the oceans amongst Royal Navy officers. As a result of his wartime experience Edgell now agreed with senior civilian scientists—such as University of Liverpool Professor of Oceanography, and fellow member of the oceanography sub-committee, Joseph Proudman—that physical oceanography was in need of greater investment, both within and outside the military research establishments. Although Edgell had interacted with oceanographers who had entered the Royal Navy Scientific Service during the war, the oceanography sub-committee of the Royal Society consisted, with the exception of Edgell, only of civilian scientists; he was the sole commissioned Admiralty representative. As well as Edgell and Proudman, both James Carruthers and George Deacon also served on the committee. This allowed Edgell to be in a position of influence, involved in composing post-war science policy within the Admiralty, and to inform those drafting this policy of the work of leading scientists. He wished to tighten the connections between the two communities.

Deacon also eventually rose to a prominent role in the committee, allowing these connections to materialise in the proposal to establish an institute devoted to oceanography. The opportunity came in early 1944 when the Royal Society sub-committee was busy discussing a very vague proposal for a post-war Anglo-Swedish expedition, which led to consideration of focusing oceanographic research in a new institution.[5] The expedition proposal was understood to be for a joint global expedition to survey the oceans, although this was unclear at the time. The proposal had taken a convoluted path through government to get to the sub-committee. It had been discussed initially by the Scientific Advisory Committee of the War Cabinet (SAC), whose senior members were also the President and Secretaries of the Council of the Royal Society. From here the proposal was passed to the Royal Society, where it was once again discussed, before being passed to the sub-committee on oceanography.

Now Edgell and Proudman used the expedition as a lever to argue for the formation of a national oceanography institute, claiming that while the Swedes could count on their own establishment to carry out the negotiations over the shape of the expedition, this was not the case for Britain as no counterpart existed.[6] This positioning now allowed the oceanographers to represent their viewpoint to both Navy and government officers as well as to discuss the proposals amongst themselves.

Although only seven pages long, the "Report of the Sub-Committee for Oceanography", published in November 1944, gave leading oceanographers like Deacon and Carruthers the opportunity to inform the Royal Society's discussions on the planning of post-war science. The report highlighted the oceanographers' wish to take a conservative stance, arguing against Hans Pettersson's proposed Anglo-Swedish oceanographic expedition and venting the fear that it would be 'parochial in its outlook and programme'.[7] This position was taken because the expedition was likely to focus on pre-war schemes rather that the emerging wartime research demands, in which neutral Sweden had not participated. Secondly these new priorities required studies of specific locales, not global observations. Yet the report recognised that the 'known interest of other nations should not be ignored' and that such an undertaking would be an excellent venture *only if* a new National Oceanographical Institute could take responsibility for it.[8] This had been the committee's ultimate objective from the outset: an institute could always press for future expeditions, whereas an expedition was unlikely to result in the creation of an institute. Furthermore an institute could solve several obstacles that oceanographers felt had blighted their scientific endeavours during the interwar period: a lack of national coordination, difficulties in gaining consistent funding, and concerns regarding the need for institutional prestige.

The bulk of the report, written by Edgell, Produman, Deacon, and Carruthers, outlined in some detail the concept of a dedicated oceanographic institute. Although there had been oceanographic institutions in Britain before the war, these had always been either based at universities (such as Liverpool and Cambridge) or under the control of government departments. Their specific aims had been dictated by a particular—at times local—research agenda, such as the MAFF Fisheries Laboratory at Lowestoft. The report detailed location, facilities, vessels, research programme, staff, salaries, and financial upkeep requirements, and governing body composition. It argued that the institute should be free of outside direct control, and be in a position to carry the research programme forward, as defined by scientists.

In a reversal of the pre-war focus on fisheries research, the Royal Society report placed the priority firmly on the physical rather than the biological side of oceanographic research. This was reflected in the suggested composition of the scientific staff for the institute: director, chief chemist, chief physicist, marine biologist, meteorologist, and geophysicist. The appendices confirmed that 'the Institute shall be primarily concerned with physical oceanography, i.e. with the dynamics, thermodynamics, physics and chemistry of the oceans and the seas. This includes a study of tides, waves, currents, circulation, temperatures, and the distribution of chemical elements in the oceans and seas, and will involve marine meteorology to some extent.'[9] The report also envisaged that the institute would 'collaborate with the Hydrographic and the D.S.R. Departments of the Admiralty'.[10] Fearing, however, that too much emphasis on the military dimensions of oceanography would have led the government to suggest funding the new institute entirely with Navy funds, the report was careful to highlight that the research carried out at the institute would stay in the "civilian" domain.

While engineering the attempt to establish a national institute, Edgell, Deacon and the other committee members needed to make sure that the Admiralty would not be made financially responsible in full for it, since mixed views existed within its ranks about investment in military oceanography. They thought about ways in which this "parcel" of financial commitment could be "passed on" to other government departments. An Admiralty memo had concluded that 'the report gives the impression that, apart from some small incidental benefit in connection with fisheries the result would be the accumulation of pure knowledge with no particular application'.[11] The Director of Finance of the Admiralty responded to the memo on 22 December 1944 stating that 'there will be no question of saddling the Navy Votes with any considerable charge'.[12] This negative response might have arisen from a misinterpretation of the research objectives of the programme, seeing proposals as being focused on biological rather than physical oceanography. It is equally likely that in defining the research programme in broad terms, Edgell may have failed to link oceanographic research to problems with ASDIC or amphibious warfare—two of the significant uses of oceanography during the war—or to emphasise the perceived and real deficiencies in British oceanographic knowledge. In part this was because the report was intended for a civilian audience as well as a military one. Edgell also had to counter opposition from non-physical oceanographers, especially fisheries scientists who wished to return to their pre-war work. Striking the right balance was difficult for Edgell: on the one hand there could be a return of oceanographic research to pre-war norms linked to commercial interests (fishing and whaling); on the other, the priority could be on continuing with wartime applications of oceanographic knowledge. The response of the Director of Finance is the last mention of Edgell's proposal that is recorded within the wartime Admiralty records, so it is unclear how the confrontation unfolded.

By 1944, officials throughout the Admiralty were looking to plan their post-war activities. For oceanography, they were looking for the right proposal, phrased in the right way, which seemed to offer the most benefit for minimal outlay. This was not the outcome that Edgell had hoped to achieve, but he still hoped to gain Admiralty support for ocean science in peacetime. Only through discussion at the SAC would the report broaden out over the last months of the war, so that when it was re-presented to the Admiralty and other government departments in late 1945, the response was at first more constructive and positive. In the process, a report written for a scientific audience was transformed by senior scientific figures into something that, when re-presented in an altered form intended for non-scientific civil servants, was viewed quite differently by figures in the Admiralty.

The main link between the Royal Society, which sought to represent the rather nebulous views of British scientists, and the wartime government, was the Scientific Advisory Committee of the War Cabinet (SAC). The SAC was made up of senior figures within the Royal Society: Sir Henry Dale, President of the Royal Society, who sat as chairman; Sir Alfred

Egerton and Professor A.V. Hill,[13] who were, respectively, the Physical and Biological Secretaries of the Royal Society; Sir Edward Appleton, the Secretary of the Department of Scientific and Industrial Research; and a representative of the civil service, John C. F. Fryer.[14] The report of Edgell's oceanography committee was first discussed at an SAC meeting on 21 December.[15] Edgell and Proudman were also present for this meeting. This was not the first time that Edgell's report had been seen by Dale, Egerton, or Hill, as the report had already been presented to the Council of the Royal Society on 2 November, and it was the Council that had forwarded it to the SAC.[16] The purpose of having the report read and discussed at the SAC was to present the proposal to the government from within the Whitehall machinery, rather than through an external lobbying body. At this meeting the report was the main agenda item, and the committee discussed almost every aspect in detail. Hill and Egerton stated at the outset that the Royal Society's Council had decided to print the report as one of the appendices, alongside other disciplinary sub-reports, to their general report on the post-war needs of fundamental research. This suggests that at least half of the committee were firmly in favour of its contents before the meeting was even held.

Nevertheless the discussion allowed further details of the report to be examined. Proudman was keen to assert the primacy of Liverpool as the most suitable location for the institute, pointing out that the university was 'anxious to collaborate in every way…and would be pleased to provide any facilities'.[17] Before the war Proudman had re-orientated the Department of Oceanography of the University of Liverpool back towards physical oceanography and had invested his whole career in the establishment of a separate Tidal Institute: he was hoping to gain significantly from the establishment of a new institute.[18] He was also keen to stress the importance of physical oceanography for 'increasing our knowledge about the movements of tides and ocean currents' and for expanding the understanding of meteorology at sea, a stance which reflected the existing programmes of the two Liverpool-based research centres.[19] Proudman was attempting to position his own agenda at the forefront of any future scheme, as there was significant potential for self-elevation for himself and for Liverpool as a centre of oceanographic research.

The SAC meeting also allowed for further discussion regarding the relationship between the proposed institute and the Admiralty. There was no precedent for the Admiralty to fund scientific research without directly managing it. Additionally there was reluctance from the Treasury to fund new research that would duplicate existing programmes. In theory strictly defined research boundaries avoided the potential for overlapping work being carried out by two establishments. But this did not fit scientific practice in reality and similar work was often duplicated as part of the testing and replication of results found elsewhere. The Admiralty and the physical oceanographers did, however, have related research priorities.[20] The study of the temperature gradients of the sea would play a significant part in any physical-oceanographic research programme, and would also be of interest to the Admiralty through their continued work on ASDIC. This was also one of the main areas of overlap between civilian and military research identified in the United States at around the same time as it was being discussed in London.[21]

The possibility of a research vessel being shared between the Admiralty and the proposed institute was also now explored in order to strengthen Edgell's original proposition. The infrastructure needs of the institute were clearly going to be both large and costly. These early discussions did not signify the birth of closely aligned ties between a civilian research institute and military research. In reality, the dynamics of civilian and military relationships, and how they would work within this scheme, remained very vague.

Over the next year the SAC considered the oceanographers' proposal at every meeting, with many of the debates focusing on the position of the institute within the government structure, or on the proposed research programme. At the meeting of 12 February 1945, the committee decided to follow up on the suggestion that the Admiralty and Institute share a research vessel; they therefore invited Charles Seymour Wright, the Director of Scientific Research at the Admiralty, to attend the meeting. Wright argued that 'certain subjects, such as the effect of noises in the sea upon the working of ASDICs, would be of little interest to the oceanographers'.[22] A shared vessel was an alternative to using Hydrographic Office vessels; at the time there was concern in the Admiralty that a dedicated oceanographic vessel would not be used all year around, and that maintenance of this vessel would fall on the Navy. Wright also remained sceptical about the potential of collaboration with civilian oceanographers, since in his opinion the differing research interests of the two parties would inevitably result in discord, which would satisfy nobody. However Wright went on to argue that 'it would, of course, be possible for the proposed Admiralty research vessel to carry out Oceanographical work on the voyages which it undertook for the purposes of his department in the same way as the Oceanographical vessel...could supply his department with useful data during their work in other fields'.[23]

So, although sceptical, Wright was clearly trying to support Edgell in his quest to prioritise physical oceanography, without financially committing the Admiralty to it.[24] However, the SAC now concluded that if a new institution was going to be established then it would have its own research vessel so as to not encroach on Admiralty's research programmes. The potential dynamic of any military-scientific relationship that the institute might eventually foster was beginning to evolve.

The reactions produced by the report go some way to explaining why Edgell had to act tactfully in reaffirming the need for a military oceanography programme carried out by civilian scientists. There were mixed reactions within the Admiralty but also among scientists interested in ocean science. These scientists were concerned about the shift in scientific priorities that would result if the Admiralty could be persuaded to emphasise military/physical oceanography research. In these circumstances Deacon emerged as someone who could mediate between opposing factions while at the same time removing the obstacles that would make it impossible for a military oceanography programme to develop.

When the Royal Society published the report of the oceanography subcommittee, Robert Sewell FRS came forward with strong objections to the proposal. Sewell was a military doctor and had been an imperial scientist between the wars, holding the position of Director of the Zoological Survey of India between 1925 and 1935; he was a senior member of the British scientific elite. Sewell had led the John Murray Expedition to the Indian Ocean during the mid-1930s, and also served as Secretary of the International Joint Commission on Oceanography from 1948 to 1951.[25] Sewell wrote a long and damning letter to the SAC complaining bitterly about the short-sightedness of his oceanographic colleagues. His complaints were wide-ranging. He believed the proposal was constructed by physical oceanographers who were attempting to sideline their biological counterparts. He proposed that a committee on which marine biologists were suitably and amply represented should reconsider the whole scheme.[26] His main criticisms were that the institution would be directed towards the 'physico-chemical side of oceanography' [sic], and the parochialism not only of working with just one nation but also of studying the Atlantic Ocean exclusively.[27] Sewell argued that it would be more productive if there could be an intensive study of different areas, carried out concurrently by several research vessels.[28]

Sewell, whose oceanographic research experience had been in the Indian Ocean, argued that of all of the seven seas, the Indian Ocean was the most justifiably ***mare nostrum*** for the British Empire and therefore it should be central to any future plans for research within a "national" institute.[29] When Sewell penned his objections, Britain was still at war with troops fighting in South-east Asia against the Japanese, and de-colonisation seemed a distant prospect to those outside diplomatic circles. In military circles, there was a strategic need for scientific knowledge of the geographical areas from where direct attacks might be launched against the British Isles or the British merchant and naval fleets, but many civilian British scientists had spent more time studying the natural environments of the colonies, rather than Western Europe or the North Atlantic. Before the Second World War, the imperial outposts still held greater importance for a number of key players in British political and scientific circles.[30] So Sewell's views were unsurprising.[31] However, any notion of an imperial imaginary for oceanography was already outdated and of little interest among the community of younger oceanographers.

Sewell failed to make an impression on his colleagues on the SAC, however, partly because the push towards a more conservative colonial policy from within the Colonial Office impacted upon the interests that he sought to defend. At the August meeting of the SAC, Professor A.V. Hill noted that the Colonial Office was keen to be relieved of its responsibilities with regard to the Discovery Committee, ruling out the possibility of financing wide-ranging oceanographic endeavours far from home. Hill therefore suggested that consideration should be given to placing the Discovery Committee under the new institute's structures of governance. Edgell, who was also present, strongly advised that any Institute vessel must not become part of the Discovery Fleet for polar investigations, but there were no other objections to the proposal to place all major British oceanographic activities under one centralised body. As a result, it was decided that the proposal, including this addition, would be taken directly to the incoming deputy prime minister Herbert Morrison, who held the post of Lord President of the Privy Council following the Labour Party victory in the general elections of July.[32] The proposal was then distributed to various government departments for their comments. This should have brought deliberations to a close as by this point the SAC had spent nearly a year considering every aspect of the proposal at length.

Sewell was still unhappy and felt that biological aspects of ocean science, particularly zoological and fishery approaches, were being neglected. He composed another memo for the committee in November 1945. This memo went into considerable detail, and the tone had now changed.

Sewell took issue with the report's use and understanding of the term "oceanography". He argued that as the proposal stood, the institute should be simply called the Hydrographic Institute, for that was what he felt the proposed research programme amounted to. He quoted remarks made by Dr Stanley Kemp in 1938 at a BAAS meeting, when he was Director of the Marine Biological Association and having served as the Director of the Discovery Investigations (from 1924 to 1936).[33] Kemp had said, 'I should like at the outset to enter a protest against the use of this word (oceanography) in a narrow and restricted sense, as a synonym of Hydrographic or the physics and chemistry of seawater. It includes within its scope not only physico-chemical work, costal surveys, soundings, studies of tides and currents, which may collectively be referred to as Hydrographic, but marine zoology and botany as well, together with some parts of geology and even meteorology. It is in this broad sense that the word is understood on the continent.'[34] Sewell wished to emphasise the importance of all these elements to the study of oceanography. He continued, 'Oceanography requires the collaboration of almost all of the known sciences and all are of equal importance, since the causation and explanation of changes in one branch can only be traced by studies in other branches of knowledge.'[35]

In Sewell's opinion the branches of science which were significant to the current study of Oceanography were 'i. Chemistry and Physico-Chemistry, ii. Marine Zoology, iii. Marine Botany, iv. Marine Bacteriology, v. Geology and Mineralogy, vi. Palaeontology, vii. Geophysics, viii. Radiology.'[36] The note then continued to explain in detail why each was important and how it interrelated with other oceanographic sciences, oceanography generally, and scientific enquiries which were not primarily based on the oceans. When circulating this note amongst the members of the SAC the new Secretary, Martin Teall Flett of the Cabinet Office staff, added his own note stating that he had informed Sewell that the oceanography proposal had been forwarded to Herbert Morrison and had stressed the importance of biological work to the institute.[37] Ultimately the second memorandum from Sewell was not discussed until the SAC met in January 1946.[38] Nevertheless Sewell was determined to challenge the physical oceanographers and eventually arranged a meeting, at the Linnaean Society, traditionally home to naturalists. Sewell viewed the Royal Society as sympathetic to the physical oceanographers' case: the Linnaean might allow the biological side to receive a sympathetic hearing.

The meeting, held on 21 March 1946, was chaired by Henry Maurice, President of the Zoological Society of London. The biological side was well represented and had a sympathetic audience in the Linnaean membership. Sewell was unable to attend as he was away in India. The only physical oceanographer invited to speak was George Deacon, but Carruthers and Edgell supported him in the discussion. Also present at the meeting was ICES president Dr Johan Hjort, a Norwegian oceanographer. In his opening remarks he stated: 'not only are all the branches of the whole study of oceanography, physical, chemical and biological interdependent, but, in combination, they minister to the safety and welfare of the nation.'[39]

Once again Deacon was able to turn the discussion so as to convince attendees to endorse his views. This was essentially the moment for Deacon to appear as a suitable mediator who could successfully represent the wishes of the Admiralty and the Navy Hydrographer without compromising support from the likes of Sewell. So far, Deacon had not played a particularly prominent role in the debates about the future of oceanography, working "in the dark" beside Edgell and Proudman; however, he was now ready to take centre stage in the debate and offer a viewpoint amenable to several groups with different agendas. These diplomatic skills, on top of his scientific ones, served a variety of purposes and would ultimately help to reinforce the network of organisations interested in establishing military oceanography.

Deacon delivered his paper first. In it he outlined in some detail the differences between the discipline of physical oceanography in Britain and in other countries, in particular in Scandinavia. However his paper quickly moved towards considerations about the future of the discipline:

> There are several indications that the study of oceanography may be entering a period in which it will receive as much, or more, support than it did after the last war. It must now be realized more than ever before that our future is intimately bound up with the sea, and the plans which have been put forward for the extension of its study should find the support they need. It is also inevitable after such prolonged and varied operations that naval authorities show more interest in the subject.[40]

Deacon did not challenge the biological oceanographers directly (if anything, he showed them new funding opportunities), and they did not disagree with him. Deacon's position could now prevail because he did not openly challenge Sewell's allies but sought a compromise between opposing factions. The main outcome was evidence of there being fewer physical oceanographers in Britain than their biological colleagues. The Linnaean meeting did not bring consensus, but it did reflect that a new generation of British oceanographers, like George Deacon, had a radically different concept of the future of their discipline than the more elderly fisheries bureaucrats, such as Maurice and Hjort. The proposal was now in the hands of the government. The SAC did not discuss Sewell's latest objection at their meeting in November 1945: plans were now in motion and beyond objections from the likes of Sewell.

Second Round (1946–48): Government Approval and Budgeting for Ocean Science

Deacon's proposed compromise appeared to offer a way forward at a time when the Admiralty, government departments, and individual civil servants had differing views about the proposed institute. The oceanographers' plans would probably never have received support, however, if it was not for the action of a new science administrator who accepted the benefits to be derived from investing in military oceanography—Henry Tizard, who was appointed Chief Scientific Adviser to the Ministry of Defence in 1947.

The reasons why the SAC had sent the proposal to Morrison were procedural. The intention of transferring a whole body, the Discovery Committee, from one ministerial responsibility (the Colonial Office), to another (the Admiralty), exceeded the considerable power of the permanent undersecretaries and the SAC Secretary Martin Flett. Flett had been the senior civil servant on the committee from the Cabinet Office, and he now approached Morrison, to whom the matter fell. Flett's memo was specifically about the future of the Discovery Committee but this was already well intertwined with the NIO proposals.[41]

This happened at a time of change for the British government and few ministers understood how to deal with the issue. The Labour Party, which came to power in a general election in July 1945 (with Clement Attlee replacing Winston Churchill as Prime Minister), had spent much of its time and energy as part of the National Government since 1931 planning for economic and social change if they did come to power. However, they appeared apathetic towards non-industrial research and a science policy for basic research was not on the pre-election policy agenda.[42] Morrison was certainly not invested in the outcome of the NIO proposal. Responding to Flett's memo, Morrison added a handwritten note:

> I am really worried about the functioning of the m/c [machinery] of govt. [government] in relation to science. This matter is scattered all over the place; I have no real knowledge as to whether or not overlapping of work is taking place; I have not the least idea whether I should back £50,000 a year for this work; nor have I adequate advice about persons. In short I feel helpless in discharging my functions despite the admirable help I get from M. Flett. I need a small scientific secretariat with its feet on the ground and probably in the Cab. [Cabinet] Secretariat and available to all depts.[43]

Morrison's lack of advice and oversight slowed proceedings. In a letter to the Colonial Office in November 1945 Sir Alan Barlow, the Second Secretary at the Treasury, outlined the financial difficulties the proposals created. Barlow was not unsympathetic to science but like many in the Treasury was resistant to funding science from the public purse.[44] Barlow's letter stated that 'there are a good many claimants of this sort for Exchequer benevolence and our first task is to assign this one to its proper place in the list. Then, if it is decided that this case is a good one, we must make recommendations on the best method of procedure.'[45] Arranging the financial accountability of the proposed institute caused more dispute and disruption to the scheme alongside peripheral debates surrounding location, programme, and personnel.

Meanwhile, Admiralty officials had become somewhat more sympathetic to the establishment of an institute, although resistance persisted about who should pay. When in early 1946, senior policy figures in the Military Branch (concerned with the operational capabilities of the Royal Navy) looked at the proposal for an institute again, they agreed to 'press for control of the administration of the new National Institute of Oceanography'.[46] The Military Branch's conviction of the need to do this led the Head of the Military Branch to suggest that it was time to take the proposal to the "board" or the Lords of the Admiralty.[47] This prompted the Under-Secretary of State within the Admiralty to intervene, responding to the Head of Military Branch, 'I confess to some doubt whether we should not do better – by which I mean get *something* for *nothing* – confining ourselves to saying in general terms that we thoroughly agree that an Oceanographic Institute is much required…how far the expenditure would be proper to Naval funds could only be judged when we know more about its programme.'[48] This negativity led others to comment: 'I should prefer to postpone our claim to responsibility for the Institute and its research until we can afford to assert it and to be content meanwhile to use the Institute for the research which we require.'[49] This stalling by senior Admiralty civil servants was kept largely within Admiralty circles, and delaying tactics seen in the correspondence between the Admiralty and other government departments would become the cause of much frustration in the Treasury and amongst oceanographers.[50]

This was a game of bureaucratic politics, playing a dual diversion of "wait and see" and "pass the parcel", and hoping the problem would resolve itself, whilst also delaying another ministry from forcing the Admiralty into making a decision. Unsurprisingly many within the Admiralty wanted a new oceanography institution to carry out research of operational significance without being made to pay for it from within their existing budget. In essence, they wished to get something (research) for nothing (no cost).

The reluctance of the Admiralty's civil servants to commit fully to funding the NIO left the scheme in a state of suspended animation, as they slowly exchanged letters back and forth with the Treasury.[51] Additionally the SAC found itself being wound down—its wartime role now long expired—prompting discussion about the creation of a peacetime arrangement that would separate civilian and defence science research policy. With the scheme essentially stalled, it took outside factors to re-ignite the proposal for an oceanography institute within the government and the Admiralty. This pressure came from Henry Tizard.

Tizard had a long history of defence and civilian research and more importantly of the administration of both these dimensions of government science. He had been involved in military research since the First World War when he had worked in the Royal Flying Corps. In 1920 he joined the newly created Department for Scientific and Industrial Research (DSIR), where he developed the skills required to be a scientific administrator, leaving as its Permanent Secretary (senior civil servant) in 1929 having established what would become the National Chemical Laboratory. Despite leaving government employment to take up a position at Imperial College, Tizard continued defence research, working on air defence projects during the 1930s and becoming Director of Scientific Research in the Air Ministry from 1934. In 1944 the chiefs of staff made Tizard the chairman of a committee established to write a report on the probable effects of new weapons on future defence policy. When the report was delivered the Labour government quickly recruited Tizard to advise on government science policy. In 1946 he acted as the chairman of the Commonwealth conference on defence science. Leading on from this, in 1947 he was appointed chairman of two committees established to resolve the needs of a peacetime SAC: the Defence Research Policy Committee (DPRC) and the Advisory Committee on Science Policy (ACSP). He also took up the role of Chief Scientific Adviser at the Ministry of Defence. This returned him to Whitehall as the most powerful scientific administrator in government service.[52]

Tizard used his chairmanship of the ACSP to connect the scientific community to government defence research policy. Working alongside Herbert Morrison and the Cabinet Office, Tizard sought to reinvigorate the proposals for an oceanography institute. It was now the oceanographers' turn to provide some additional input, but—aside from Deacon—most of them proved to be rather confused about the way forward.

Oceanographers had themselves become concerned about the lack of progress towards the formation of the NIO. University-level oceanographic programmes such as those at Cambridge had already restarted and were beginning to press for a renewed expedition programme to complement their own teaching activities.[53] On 29 May 1947 Edward Bullard, Director of the Cambridge School of Geodesy and Geophysics, wrote a letter to Henry Tizard explaining that he had been to see the Hydrographer to ask him about the likelihood of there being an available oceanographic research ship.[54] Bullard explained that the Hydrographer had stated that there were differences of opinion between the Admiralty and other departments and that he did not expect rapid progress. Bullard himself had differences of opinion with certain government departments. As he put it, 'it really does seem to me a misuse of resources, to use the only ship we have with a deep sea winch for taking rations to lighthouses or cutting the wicks of the lights on buoys, or whatever it is that Trinity House use the Discovery for'.[55] Tizard acted quickly, asking the Secretary of the ACSP and DRPC, Ian Montgomery, a civil servant in the MoD, to draft a reply to Bullard.[56]

Instead of replying to Bullard, Montgomery wrote to Neil Alison Mackintosh, Director of Research for the Discovery Committee, requesting him to advise 'in confidence what the full facts of the situation are' so as to ascertain whether it was now time for Tizard or ACSP to intervene.[57] Mackintosh had been a longstanding member of the Discovery Committee, developing techniques for whale marking in the Southern Ocean during the 1930s, but by remaining in Cambridge for the duration of the war he had kept outside military research structures. Mackintosh's response reflected his lack of active engagement with bureaucratic wartime structures: 'I am afraid it is rather difficult for me to give a helpful answer…since I have not personally been concerned with all the details of recent discussions between the Colonial Office and other Departments about the Discovery Committee's future …I should be very glad to give any assistance I can, but hesitate to give an opinion without reference to the Colonial Office.'[58] Mackintosh had left himself out of most of the planning for the future of the Discovery Committee, seeing his role as being to shape the research programme rather than finding funding for the research.[59] He left matters with the Secretary of the Discovery Committee, who also had to hound Trinity House for the return of the RRS *Discovery II*.[60] Even though the future of the Discovery Committee had been under discussion at government level for a number of years, Mackintosh seemed ambivalent. Tizard responded to Bullard informing him that 'the Government has decided to encourage the study of oceanography in a much more comprehensive way', thus making it clear that the intention of the new Labour Cabinet was to sponsor the establishment of a new institute rather than several separate research endeavours in oceanography.[61] Three years after the initial proposal, the only plan that had been suggested and agreed was that some form of action would be taken. With the chair of ACSP/DPRC on board, it was now less likely that Treasury officials would object to funding a new institute. Yet, they would place emphasis on Admiralty funding while the Admiralty would expect the Treasury to open the purse.

In the immediate post-war years the Royal Navy underwent a period of severe budget cuts designed to reduce the impact of defence spending on the government budget.[62] In addition, British strategists and science advisers like Tizard believed that the airborne nuclear weapon platform would direct future war plans. Until the Royal Navy could develop a rival nuclear weapons system, it was to an extent no longer the "senior service" within the British armed forces. Those at the helm of the Admiralty were well aware of this, knowing that the RAF would fight to control spending on a seaborne nuclear deterrent, so any research designed to retake the nuclear initiative from the RAF would need to be done from within a severely restricted funding regime.[63] This is why the initial proposal to place the bulk of the funding for the NIO at the Royal Navy's feet was unwelcome within the Admiralty. A research programme that was designed to produce a greater understanding of the workings of the seas without appearing to produce immediate benefits to the Royal Navy, however admirable, was unlikely to appeal to the conservative Lords Commissioners of the Admiralty.

What support existed within the Admiralty for the NIO began to melt away once again as the increasingly restrictive budget estimates began to emerge. Internal attitudes in the Admiralty were divided. Charles Blake Coxwell, Permanent Under-Secretary at the Admiralty, wrote to the Deputy Secretary of the Admiralty, Richard Powell, on the issue in July 1947. He argued that 'the man in the street, I suspect, would certainly be surprised if he were told that the Admiralty did not have the major interest in matters related to the oceans, and wish to take the lead therein; and the House of Commons would very probably take the same view.'[64] The Deputy Secretary responded that 'Barlow [of the Treasury] cannot fail to recognise the possibility that if he presses the financial responsibility upon us, it is not impossible that we may put the whole scheme back into the melting pot.'[65] In a letter then composed as an official response to Barlow and the Treasury in general, the Admiralty accepted ministerial responsibility on these grounds but failed to commit to the expenditure.[66]

The Treasury response emphasised that the time for decisions was imminent; it had been nearly two years since Herbert Morrison had provisionally approved the proposals. The Under-Secretary for Finance within the Admiralty had been conducting his own discussions with the Treasury. Explaining the benefits that would accrue to the British economy and security from such an investment, he had likened the NIO to medical and agricultural research, in an attempt to persuade the Treasury officials that funding should come from the scientific investigations vote. He reported to Charles Coxwell, Permanent Under-Secretary, that this suggestion led the Treasury to ask whether the Admiralty's 'disinclination to shoulder the financial responsibility was sufficiently pronounced' to drop or defer the scheme indefinitely.[67] Even the Under-Secretary for Finance, who had been one of the main opponents of the Admiralty funding either biological research or "pure science", now informed the Permanent Under-Secretary that the Admiralty would either need to 'shoulder full responsibility for Oceanographical Research, both ministerial and financial' or risk losing the scheme entirely. [68]

What the finance department of the Admiralty wanted to hear was that the Treasury would take financial responsibility for the programme, even if this required some contribution from the Admiralty and provided that the institute would prioritise physical oceanography. The Head of the Military Branch, the Hydrographer, and the Head of the RNSS saw this as a happy compromise, but others needed further reassurance this was achievable in practice. This stalemate was not just frustrating for Henry

Tizard and Edward Bullard; it was also irritating to members of the Scientific Service in the Admiralty.

Once again, Deacon stepped in to find a compromise. He could now exploit the greater influence that his mediation on the occasion of the Linnaean conference had given him, and his acquaintance with leading science administrators such as Frederick Brundrett helped him in this. Brundrett had started his career at the Admiralty Wireless Research Establishment in 1916–19, going on to replace Charles Seymour Wright as Chief of the Royal Navy Scientific Service in 1945. He knew the intricacies of defence research management extremely well and was well placed to support those members of the RNSS, such as Deacon, that he wished to endorse.[69] Thanks to Brundrett's support, Deacon was made responsible for reconfiguring the proposal for an institute so as to make it responsive to the agenda of the new government and the Admiralty, while at the same time appeasing the scientists who had been opposed to a programme focusing on physical oceanography.

Deacon understood that the chief bone of contention was one of emphasis rather than substance; biological research ought to find a place in the planning but should not compromise the orientation towards the physical that the Admiralty wanted. So he now worked towards completing a memo (see Fig. 3.1) that presented a restructured vision of the institute's proposed research programme so as to emphasise the "military" usefulness of the institute. Deacon's proposal placed biological research tenth, just before the work of the Discovery Committee at eleventh, on an eleven-point proposal. Deacon's proposal was written purely for the Admiralty; it was not intended for oceanographers outside the Admiralty research community. Despite its lowly position on his ranking, he was still able to show that there was a military interest in the fouling of ships by marine organisms. The bulk of his proposals had to do with the operations of submarines and the propagation of sound through water. This was what Charles Seymour Wright had argued that oceanographers would not be interested in back at an SAC meeting in early 1945. The memorandum played an important part in shifting opinion within the Admiralty towards the institute.

Following Deacon's first memorandum that had altered the order of priority for research, Brundrett asked him to compose a second that gave further details on costs.[70] The memo explained that since 1945 the needs of the Royal Navy had resulted in the growth of staff and facilities based at the Admiralty Research Laboratory (Teddington). This research was

Memorandum for C.R.N.S.S.

Probable Research Programme of a National Institute of Oceanography
governed by a by a body composed of representatives of Government offices, the learned societies and the Universities

	General Study	Interest to the Admiralty
1	Bottom configuration and sediments.	Applications to submarine detection and evasive tactics of submarines; and to some mining problems.
2	Physical properties of sea water	Information on penetration of light and transmission of sound
3	Distribution of Physical Properties	Information on temperature salinity and density gradients which influence the effective range of underwater detection, evasive tactics and trimming of submarines. A study of the occurrence and causes of natural electric potential gradients, and fluctuation of magnetic field, in the sea is needed for the development of new methods of underwater detection and weapons.
4	Ocean currents, wind currents and tidal streams	This study is fundamental to a knowledge and predication of the water conditions which influence submarine detection and evasive tactics. More information is required for air-sea rescue work.
5	Waves, swell and surf	Information is needed (1) to improve prediction of waves, swell and surf from meteorological charts or from anticipatory measurements. (2) for a study of the factors that influence landing craft, and the design and position of breakwaters and harbours. (3) of the pressure changes and microseismic ground movements associated with waves, in connection with development of methods of underwater detection and mining.
6	Internal waves in the sea	Information is required of their influences on methods of underwater detection and changes in trim of a submarine
7	Interaction between the sea and the air	The heat exchanges near to the surface influence the effective range of radar, and the behaviour of smoke screens
8	Chemistry of Sea water	Some applications to the behaviour of materials used under the sea
9	Instruments	There are Admiralty requirements for new types of tide recorders, current meters and temperature and salinity recorders.
10	Biology. Although this Institute would not compete in biological studies with the existing Marine Biological and Fisheries Laboratories it would be expected to assist them whenever possible, and to be aware of the biological significance of its physical studies	The Admiralty's interest is confined mainly to information on fouling, boring and noise-making organisms.
11	Work of the Discovery Committee (The Position of the Falkland Island Dependencies Survey – founded largely for political reasons – needs to be defined).	Although mainly biological, the observations in the Southern Ocean make a large contribution to the solution of physical problems.

GERD, A.R.L. TEDDINGTON, 6 October 1947

Fig. 3.1 Deacon's memorandum: proposed NIO research objectives (1947)

costing £25,000 per annum. Deacon explained that future demands required the acquisition of information pertaining to the regional and seasonal changes in oceanographical conditions, which influenced methods of underwater detection and the operation of a submarine.[71] Deacon had stated at the end of his first memo that 'the institute would have no provision for applying the results of its fundamental researches to specific Admiralty requirements of a secret nature, but it would possess the facilities necessary for Admiralty scientists to learn and investigate the aspects of oceanography most closely associated with such requirements'.[72] What constituted 'Admiralty requirements of a secret nature' Deacon did not specify, but its inclusion would have been clear to its intended audience amongst the civil servants of the Admiralty: the collection of oceanographic information in principle, but its applications to anti-submarine warfare as well.

Deacon also capitalised on his ability to persuade friends and rivals alike by asking to review the original plans, a process that revealed that the views of Proudman no longer represented the wider feelings of British oceanographers in these debates. Deacon understood that Admiralty requirements had to be at the heart of the institute's priorities, but he rejected the original plan to locate the institute in Liverpool—in contrast with Proudman's early plans. He argued that the enlarged scope of the original proposals, as presented in 1944, meant that Liverpool was no longer a good choice for the location of the institute. Instead Deacon recommended that a much larger site, closer to the 'Admiralty establishments' would be most suitable for the institute.[73]

On the 11 February 1948, a meeting was finally held which brought together the three organisations that would make up the NIO. Present were the Head of the Military Branch (Nigel Abercrombie), the Hydrographer (Rear Admiral Arthur Norris Wyatt), and the Chief of the Royal Navy Scientific Service (CRNSS) (Frederick Brundrett), along with Deacon, Carruthers, and Neil Mackintosh. Wyatt felt that deliberations had gone on too long and it was time to scrap the notion of an independent institute and return to Admiralty-directed oceanographic work. This could have been posturing on Wyatt's part, trying to ensure that his department was not forced to share resources, nor would his position be challenged by appointment of a "senior" oceanographer to lead the institute.[74] It seems unlikely that Wyatt was merely frustrated with all the indecision, but Brundrett, who had not been involved in the proposal for as long as Wyatt, was more enthusiastic. He argued that it was 'preferable to start on the right lines straight away; the nucleus was already available in the Admiralty and the existing Discovery Committee'.[75] Mackintosh agreed, pointing out that his committee needed to concentrate on work at home, producing greater scientific output at less cost. The meeting also approved a second phased and costed programme prepared by Deacon, although the minutes show that there had been input from Carruthers and Mackintosh. From three separate programmes came the notion of one unified picture. However, regardless of how much consensus was recorded in the minutes, the reality was somewhat different.

Five days after the meeting Carruthers wrote a furious memo. He felt that he and the war work undertaken were now being ignored: 'if translated into fact, the proposals made in the recent memorandum concerned with costing etc., could hardly fail to leave the Hydrographer's unit in a junior position in the embryo Institute.' Carruthers was not only worried about "his" department but also about his own career: 'it would seem no more than equitable if the Heads of the three component parts started off (once their unites [sic] were complemented) with equal rank, but with some sort of command pay and special standing given to him, who was chosen as Director.'[76] Additionally Brundrett now made a move in a letter to Abercrombie—he argued that Mackintosh 'has a radically different conception of the final organisation from anything I feel we can accept'.[77] Yet Brundrett took the opportunity to put forward the case for Deacon, who was the only representative at the meeting from the RNSS. Brundrett argued 'the quite universal opinion is that there is only one possible candidate for the post of Director of the Institute and that is Deacon.'[78] Since Deacon had demonstrated that he could find a compromise with other components of the new institute, he now appeared to Brundrett and others as the most suitable candidate. Having Brundrett as a supporter was to prove extremely useful for Deacon at this time and in the future.

Presenting his programme proposals had clearly done little harm to Deacon's career prospects. At an earlier meeting in which his memorandum was discussed, the minutes recorded: 'there is little doubt, in fact, that if the National Institute were set up as an independent body, Dr. Deacon would be invited to occupy a very high, if not the highest, position in it.'[79] Independently, Neil Mackintosh now sent his own memorandum, drawn up on 20 February, to the Cabinet Office. In this memo Mackintosh argued that 'it is essential however that the Discovery Investigations should be kept distinct and under separate direction from the Institute'.[80] This demonstrates that even within "consensus" there remained many unresolved disputes. Creating something acceptable to the Admiralty was one thing; suppressing the divisions amongst oceanographers was another. With Mackintosh having finally found a voice in the discussions, he now pressed forward his case for the retention of as much Discovery Committee independence as he could. At stake was the timescale and shape of the process, which would see the Discovery Investigations dovetail the work of the new institution.

It was not until July 1948 that the approval of the First Sea Lord for the creation of the NIO was obtained. In his memorandum, Abercrombie, who had become Head of the Military Branch, noted that approval had only been given on the understanding that the annual charge could be 'reasonably borne'. Abercrombie stated that he imagined the amount the Navy would have to pay out would be around £120,000 in the first year and £50,000 annually thereafter. Therefore, Abercrombie added, it is 'important that this country should not lag behind in the field of oceanographical research. The Americans and indeed, it is believed, the Swedes are already well ahead and for the sake of national prestige alone it is desirable that we should get started as soon as possible.'[81] Abercrombie mixed the fear of lagging behind allies and foes with a sense that any future war at sea would be as much a fighting war as a demonstration of scientific power. His attitude was that scientists needed to be kept on tap, rather than in reserve, and this was becoming increasingly apparent as the Cold War began to worsen, with tensions heightened with the onset of the Berlin Blockade (June 1948–May 1949).

Third Round (1949–51): Governance of the Institute

In 1948 the UK Cabinet Defence Committee stressed that anti-submarine warfare (ASW) had to become the Royal Navy's chief responsibility. This assessment was based on the accepted belief that Britain would not be in an economic position to fight a war before 1957, and that if war did come, extensive support from across the Atlantic would be needed in naval operations.[82] Renunciation of a more proactive role in naval defence globally, in favour of one that the British economy could sustain, was thus inevitable. The Royal Navy's role in patrolling the Eastern Atlantic on behalf of its allies had only been secured after Churchill delivered an impassioned speech beseeching US President Harry S. Truman to 'make room for Britain to play her historical role "upon that Western sea whose floor is white with the bones of Englishmen"'.[83] However, over the next few years British military crises globally, such as the Yangtze Incident (1949), the Korean War (1950–53), and Malayan Emergency (1948–60), demonstrated that the Royal Navy still retained some influence in international sea operations. If ASW was to be the priority then the Royal Navy Scientific Service needed to be orientated to this new goal; the creation of the NIO was partly a response to this bringing expertise together under one organisation.

The establishment of the NIO brought together three bodies—the civilians in the Hydrographic Office, George Deacon's W Group, and Mackintosh's Discovery Committee. The final part of the process of creating the NIO was the building of structures of governance around the Institute, which would by extension have wider implications for oceanography in Britain. This final stage allowed questions of the Institute's location to be addressed, along with staffing and a return to the question of expedition programmes and objectives.

In 1948 a Royal Charter was passed which brought the NIO into existence (if only on paper), as of 1 April 1949. The charter gave the Institute a guaranteed foundation by law. Due to the Institute's Royal Charter status, it was protected from being closed or disbanded by successive governments, but this did not dictate the amount of funding that should be allocated, or from where these funds should come. One of the key aspects of the charter was its ratification of the National Oceanography Council (NOC), a government body that would oversee the work of the Institute and be able to voice opinions on its programme as set out by its director. However the Institute had little else. There were no physical premises, there were no ships (although the Discovery Committee vessels were quickly transferred under the Institute's control), and there was no immediate programme. Reviewing all of these administrative problems became the first task of the NOC in 1949.

The exact composition of the NOC had changed several times in the period leading up to the passing of the Royal Charter, and its establishment was central to continuing military–civilian relationships. For the Admiralty, the acceptance of the NOC as a body that could shape policy at a research establishment, which they funded, was a major departure. Traditionally the Admiralty would have placed a naval officer in charge of an establishment to establish primary governance, with his superiors in a strict hierarchical regime. In reality the day-to-day running of the NIO was to be left in the hands of its director, with a governing body outside the Admiralty. This meant that from a strictly local perspective Deacon was the only "node" that could feed into the network the policy positions of the Admiralty. The NOC, made up of representatives from throughout the scientific and Whitehall communities, played a significant part in maintaining the NIO's position as a civilian entity (see Fig. 3.2).

<u>Members of the National Oceanographic Council</u>

Name	*Office-Appointment-Department*
Mr. W. R. J. COOK, C.B., M.SC. (Vice-Chairman)	Chief of the Royal Navy Scientific Staff
Rear-Admiral A. DAY,	Hydrographer
Mr. S. GRAHAM SMITH,	Assistant Secretary to the Admiralty
Mr. J. J. S GARNER,	Deputy Under-Secretary of State for Commonwealth Relations, Commonwealth Relations Office
Mr. C. G. EASTWOOD,	Assistant undersecretary, Colonial Office
Mr G. M. GRAHAM,	MAFF
Mr. H. J. JOHNS,	MAFF
Mr. A. J. AGLEN	Scottish Office
Dr. C. E. LUCAS	Director of Marine Laboratory, Fisheries Division, Scottish Home Department
Sir NELSON JOHNSON	Director Meteorological Office
Dr. E. C. BULLARD	Head of National Physical Laboratory
Dr. H. W. PARKER,	Director of the British Museum (Natural History)
Dr. F. W. G. WHITE	Chief Executive, Commonwealth Scientific and Industrial Research Organisation (Australia)
Dr. E. MARSDEN	New Zealand Government, Scientific Adviser in London
Mr. E. R. A. DE ZYLWA	Director of Fisheries for Ceylon
Sir DAVID BRUNT	Vice-President of the Royal Society
Dr. A. T. DOODSON	Director of the Tidal Institute (Liverpool)
Sir GEOFFREY TAYLOR	Yarrow Research Professor, Cavendish Laboratory (Cambridge)
Professor W. B. R. KING	Professor of Geology at University of Cambridge
Mr. J. M. WORDIE	President of Royal Geographical Society
Professor H. JEFFREYS	Director of the International Seismological Summary
Dr. J. B. TAIT	Marine Laboratory, Aberdeen
Mr. F. S. RUSSEL	Directory Marine Biological Association (Plymouth)
Dr. D. D. JOHN	Ex-Member of the Discovery Investigations
Professor C. M. YONGE	President of the Scottish Marine Biological Association
Professor A. G. OGILVIE	President Scottish Geographical Society
Professor J. PROUDMAN	Professor of Oceanography University of Liverpool
Professor G. R. GOLDSBROUGH	Retired Professor of Mathematics, Kings College, Newcastle
Professor J. RITCHIE	Vice-President Royal Society of Edinburgh
Mr. B. C. BROWNE	Head of the Department of Geodesy and Geophysics, Cambridge

Fig. 3.2 Members of the National Oceanographic Council, 1951

Admiralty personnel had initially shown a degree of scepticism towards the composition of the NOC. They felt that the body was too large and cumbersome, while others such as Tizard and Deacon argued precisely for this large composition. The NOC was a key factor in making the NIO a "national" centre for oceanography. It knitted the NIO into the web of disparate oceanographic establishments around the country, as well as into the major scientific societies. Governed by representatives of all the other groups, the NIO had a national identity and was a central focal point for British oceanography. There were strong objections amongst members of the Royal Society of Edinburgh that the Labour government had neglected to make national bodies anything other than English and London-centric. These objections contributed to ongoing debates regarding the eventual location of the NIO. The NOC itself represented almost every geographical location around the UK, especially among its scientific council members, and through the university and scientific institutional representatives.

Across the UK the discipline was highly fragmented: individuals taught courses on oceanography at various universities, and prestigious bodies such as the Royal Society, Royal Geographic Society, and the Discovery Committee were undertaking large international expeditions. As the CRNSS, Charles Seymour Wright, put matters in 1945, 'There is nothing in the UK that can be compared with the U.S. Oceanographical institutes...nor to the German Deutsche Seewarte. We are, I think, less well informed on these matters than Norway, Sweden, Holland and possibly Denmark and France.'[84] This lack of a central focus had been significant during the war. In the United States, the Navy had gone straight to the major oceanography centres (Woods Hole Oceanographic Institute, Lamont, or Scripps). In the UK, instead of having a central pool of expertise, oceanographers had been swallowed up into Admiralty research establishments, left largely to their own devices, or played a minor role in the Committees of the Royal Society. This also meant that when the sub-committee report was delivered to the Royal Society and subsequently passed on to the government through the auspices of the Scientific Advisory Committee, many of the most "active" oceanographers had been left without a say in the future both of their discipline and of their own careers.

The report of the Royal Society's sub-committee for oceanography had nominated Liverpool, and more specifically the University of Liverpool, as the ideal home for the National Institute of Oceanography. This was unsurprising, considering that a major contributor to the report was

Liverpool's Professor of Oceanography, Joseph Proudman. However the city did have a strong claim for hosting the institution. Not only did the Port of Liverpool provide ready access to the sea and a convenient base for research vessels, the university also housed a well-established Department of Oceanography, which could provide teaching facilities and staff. The Tidal Institute was also located nearby.[85]

The question of location was one of the few parts of the report that had not been challenged during the first phase of government-level negotiations. Those within the government who were consulted about the NIO in late 1945/early 1946 who did have experience of working with universities and science more broadly did express some concern. In January 1946 the Director of the Meteorology Office, Sir Nelson Johnson, wrote to Barlow stating, 'there are several other universities and kindred establishments which are equally, or even more, interested in the project than Liverpool University. The universities of London, Edinburgh, Glasgow, Aberdeen and the Marine Biological Institute at Plymouth may be mentioned. The collaboration of the experts at these places is more likely to be secured if the proposed Oceanographic Institute is an independent establishment, in the running of which they have some say, than if it were made the responsibility of Liverpool University.'[86] If Proudman had served on the committee and all of the financial advantage of the institute was given to the University of Liverpool then there would be others who would feel aggrieved. There was a conscious effort on the part of Edgell and Proudman to ensure that the new institute would be independent from existing institutes such as the biological oceanography centres at the MAFF laboratories, Plymouth, or the Aberdeen (Torry) Fishery and Marine Laboratory. This was in part because of the perceived dominance of biological research over physical during the pre-war period, but also because the Institute was seen as being something new, which had to be free of existing institutional politics, allowing for a new more balanced approach linking physical and biological oceanography.

When Tizard and the ACSP began to send out invites to institutions and bodies to join the NOC, the question of location arose. Placing the NIO under the NOC, Sir Nelson Johnson's fears were alleviated somewhat in that Liverpool University was not to run the Institute. Few, save Proudman, favoured Liverpool. Having listened to the views of the Scottish representatives (and ignored them), the NOC came to examine the two strongest claims, those of Liverpool and London. When the NOC Executive Committee met to discuss this in 1950, Deacon believed that the merits which Liverpool had once held were now irrelevant.[87] London or at least a site near London had several advantages and was significant from the perspective of the geography of science policy, mainly the close proximity to the NOC meetings whose large and diverse membership found this a logistically convenient travel destination. It was also closer to the Admiralty research establishments on the coast. However, it would not be until 1953 that all of the component parts of the NIO scientific structure would come together under one roof in a disused building near Wormley. The building, which had once housed the Signal and Radar Research Establishment, remained a compromise. It was many miles from any actual seawater, although it did provide good rail links with Portsmouth and London. So both its military and civilian visitors could reach it fairly quickly.[88] The location became a fitting embodiment of the Institute's potential dual purpose, being midway between "militarised" Portsmouth and "scientific" London (Fig. 3.3).

Fig. 3.3 National Institute of Oceanography building, Wormley, 1953. (Image from the Archives of the National Oceanographic Library, National Oceanography Centre, Southampton)

The events surrounding the founding of the NIO resembled a complex bureaucratic game of "pass the parcel", which continued for much longer than anyone had foreseen in 1945.[89] This game was played at the bureaucratic level between the civil servants of the Admiralty and the Treasury, whilst being similarly picked up and dropped by various members of the shifting policy community throughout Whitehall as the issue surfaced, re-emerged, and eventually became pressing. Nevertheless, this "troublesome birth" did ultimately result in a prolonged period of stability for the NIO. Although it may appear as if Deacon and his vision for post-war oceanography in Britain had triumphed, his power base was only maintained with the tacit assent and somewhat grudging support of his fellow scientists. Ultimately, however, they also gained out of the creation of the Institute, finding career stability and access to resources that had previously been out of reach. Deacon succeeded because he could find a compromise while essentially endorsing the Admiralty's viewpoint in negotiations, thus successfully making the network functional.

Notes

1. Phillip J. Gummett and Geoffrey L. Price, "An Approach to the Central Planning of British Science: The Formation of the Advisory Council on Scientific Policy," *Minerva*, 15:2 (1977): 119–143.
2. Note from Deputy Secretary to Coxwell, Permanent Under-Secretary, Admiralty, 11 July 1947, ADM 116/5715, TNA (London).
3. David Edgerton, *Britain's War Machine: Weapons, Resources, and Experts in the Second World War* (Oxford: Oxford University Press, 2011): 135.
4. Edgell's final report for the Admiralty was submitted to the Admiralty post-war problems committee on 17 July 1944, ADM 1/16020, TNA (London).
5. The Royal Society Minutes of Council, 1940–45, Vol. 16, Minutes of Meeting 13 January 1944, 209.
6. Margaret Deacon, "Steps toward the founding of the NIO," in *Of Seas and Ships and Scientists: The Remarkable Story of the UK's National Institute of Oceanography*, Anthony Laughton, John Gould, M.J. 'Tom' Tucker, Howard Roe, (Cambridge: Lutterworth Press, 2010): 25–6.
7. Report of the Sub-Committee for Oceanography, 24 August 1944, CAB 124/555, TNA (London).
8. Ibid.
9. Appendix B, Report of the Sub-Committee for Oceanography, 24 August 1944, CAB 124/555, TNA (London).

10. Ibid.
11. Memo Head of Military Branch, 19 December 1944, ADM 116/5715, TNA (London).
12. Note Director of Finance to Head of Military Branch, 22 December 1944, ADM 116/5715, TNA (London); "votes" was the money given to the Navy by Parliament as their annual budget.
13. It should also be noted that A.V. Hill was heavily involved in the Marine Biological Association so knew something of marine science.
14. W.S. Feldberg, "Henry Hallet Dale, 1875–1968," *Biogr. Mems. Fell. R. Soc.*, (16:1970): 77–174; D.N. Newitt, "Alfred Charles Glyn Egerton. 1886–1959," *Biogr. Mems. Fell. R. Soc.*, (6:1960): 39–64; B. Katz, "Archibald Vivian Hill. 26 September 1886–3 June 1977," *Biogr. Mems. Fell. R. Soc.*, (24:1978): 71–149; J.A. Ratcliffe, "Edward Victor Appleton. 1892–1965," *Biogr. Mems. Fell. R. Soc.*, (12:1966): 1–21; H. Edelsten, "John Claud Fortescue Fryer. 1886–1948," *Obit. Not. Fell. R. Soc.*, (7:1950): 94–106.
15. Minutes of meeting 21 December 1944, CAB 90/5, TNA (London).
16. Royal Society Minutes of Council, vol. 16, 315.
17. Minutes of meeting 21 December 1944, CAB 90/5, TNA (London).
18. Anna Carlsson-Hyslop, "An Anatomy of Storm Surge Science at Liverpool Tidal Institute 1919–1959: Forecasting, Practices of Calculation and Patronage," (Ph.D diss., University of Manchester, 2010): 84.
19. Minutes of meeting 21 December 1944, CAB 90/5, TNA (London).
20. Ibid.
21. Hamblin, *Oceanographers and the Cold War*, p.xx.
22. Minutes of Meeting, 12 February 1945, CAB 90/6, TNA (London).
23. Minutes of Meeting, 12 February 1945, CAB 90/6, TNA (London).
24. See Peder Roberts, *The European Antarctic: Science and Strategy in Scandinavia and the British Empire* (New York: Palgrave Macmillan, 2011).
25. C.F.A. Pantin, "Robert Beresford Seymour Sewell, 1880–1964," *Biogr. Mems Fell. R. Soc.* (1965): 150, 152. The International Joint Commission on Oceanography was established by ICSU in 1948 to concentrate on the coordination of deep sea research. Torben Wolff, *The Birth and First Years of the Scientific Committee on Oceanic Research (SCOR)*, http://www.scor-int.org/Publications/SCOR_History_1.
26. Memorandum for the Council of the Royal Society in connection with the 'Report on the needs of research in fundamental sciences after the war', *Post-War Research in Oceanography*, R. B. Seymour Sewell; 3 March 1945, CAB 90/6, TNA (London).
27. Ibid.
28. Hamblin, *Oceanographers and the Cold War*, 62.

29. *Mare nostrum* meaning 'our sea'.; 'A Note on the Proposal to Establish a National Institute of Oceanography', R. B. Seymour Sewell, 22 November 1945, CAB 90/6, TNA (London).
30. See Jim Endersby, *Imperial Nature: Joseph Hooker and the Practices of Victorian Science* (Chicago: University of Chicago Press, 2008); Brett M. Bennett and Joseph M. Hodge, eds, *Science and Empire: Knowledge and Networks of Science across the British Empire, 1800–1970* (London: Palgrave Macmillan, 2011).
31. John M. MacKenzie, *Propaganda and Empire* (Manchester: Manchester University Press, 1986).
32. Minutes of SAC Meeting, 9 August 1945, CAB 90/6, TNA (London). Lord President is the fourth highest office of the British State. The Privy Council is an ancient body which links the monarch to the parliament and acts on their behalf in the examination of bills and the signing into law of legislation. They also manage bodies under Royal Charter, such as universities.
33. A.C. Hardy, "Obituary. Stanley Wells Kemp F.R.S. 1882–1945," *Journal of the Marine Biological Association of the United Kingdom,* 26 (1946): 219–234.
34. 'A Note on the Proposal to Establish a National Institute of Oceanography', R. B. Seymour Sewell, 22 November 1945, CAB 90/6, TNA (London).
35. Ibid.
36. Ibid.
37. SAC (45) 24, M.T. Flett, 22 November 1945, CAB 90/6, TNA (London).
38. SAC (46) 1st Meeting, 14 January 1946, CAB 90/6, TNA (London).
39. H.G. Maurice, "Discussion on Oceanography," *Proceedings of the Linnaean Society of London,* 158 (July 1947): 79.
40. George E.R. Deacon, "Physical and Chemical Problems in Oceanography," *Proceedings of the Linnaean Society of London,* 158:2 (July 1947): 81.
41. M. Flett, memo for Herbert Morrison, 'Discovery Committee', 22 October 1945, CAB 124/555, TNA (London).
42. On basic science, see Sabine Clarke, "Pure Science with a Practical Aim: The Meanings of Fundamental Research in Britain, circa 1916–1950," *Isis,* 101 (2010): 285–311; Ronald Kline, "Constructing 'Technology' as 'Applied Science': Public Rhetoric of Scientists and Engineers in the United States, 1880–1945," *Isis,* 86 (1995): 194–221; Jane Calvert, "What's Special about Basic Research?" *Science, Technology and Human Values,* 31 (2006):199–220; Désirée Schauz, "What is Basic Research? Insights from Historical Semantics," *Minerva,* 52 (2014): 273–328; Roger Piekle Jr., "'Basic Research' as a political symbol," *Minerva,* 50 (2012): 339–361.
43. Herbert Morrison, 23 October 1945, CAB 124/555, TNA (London).
44. Deacon, "Founding of the NIO," *Of Seas and Ships,* 30.

45. CO 927/7/6, letter from Barlow to Sir George Gater, 6 November 1945, CO 927/7/6, TNA (London).
46. Minute by Head of Military Branch, 8 February 1946, ADM 116/5715, TNA (London).
47. Ibid.
48. Minute by USS to Head of M. Br, 11 February 1946, ADM 116/5715, TNA (London). Italics mine.
49. Follows above in minute sheets, author unknown, 19 February 1946, ADM 116/5715, TNA (London).
50. Minute sheets 1946–7, ADM 116/5715, TNA (London).
51. The vote was the annual budget given to government ministries by the Treasury.
52. G.J. Piller, "Tizard, Sir Henry Thomas (1885–1959)," *Oxford Dictionary of National Biography,* (Oxford University Press, 2004) doi:10.1093/ref:odnb/36528 [accessed 14/12/14]; R.V. Jones, W.S. Farren, "Henry Thomas Tizard, 1885–1959," *Biographical Memoirs of Fellows of the Royal Society,* (1961); on the Defence Research Policy Committee, see Jon Agar, Brian Balmer, "British Scientists and the Cold War: The Defence Research Policy Committee and Information Networks, 1947–1963," *Historical Studies in the Physical and Biological Sciences,* 28:2 (1998): 209–252.
53. The NIO during the 1950s allowed Bullard the use of the RRS *Discovery II* for two months every other year.
54. Letter Edward Bullard to Henry Tizard, 29 May 1947, CAB 124/555, TNA (London).
55. Ibid.
56. Note, Henry Tizard to I. Montgomery (MoD), 4 June 1947, CAB 124/555, TNA (London). On the formation of ACSP, see Phillip J. Gummett, Geoffrey L. Price, "An Approach to the Central Planning of British Science: The Formation of the Advisory Council on Scientific Policy," *Minerva,* 15:2 (1977): 119–143; Montgomery is identified as the Secretary of ACSP and DPRC in Jeff Hughes, "Doing Diaries: David Martin, the Royal Society and Scientific London, 1947–1950," *Notes & Records of the Royal Society,* 66 (2012): 273–294, note 49.
57. Letter I. Montgomery (MoD) to Dr Mackintosh (Discovery Committee), 6 June 1947, CAB 124/555, TNA (London).
58. Letter Dr. Mackintosh to I. Montgomery, 6 June 1947, CAB 124/555, TNA (London).
59. Peder Roberts, "A Frozen Field of Dreams: Science, Strategy, and the Antarctic in Norway, Sweden and the British Empire, 1912–1952," (Ph.D diss., Stanford, 2010): 130.
60. Correspondence between S. W. Smith (Secretary of the Discovery Committee) and Miss Churchard (Ministry of Transport), CO 927/39/1, TNA (London); CO 927/39/2, TNA (London); CO 927/39/3, TNA

(London). The corporation of Trinity House has since 1514 been responsible for the maintenance of navigational aids (such as lighthouses) in British waters.

61. Letter Henry Tizard to Edward Bullard, 17 June 1947, CAB 124/555, TNA (London).
62. Eric Grove, *Vanguard to Trident: British Naval Policy since Second World War* (Annapolis, MD: Naval Institute Press, 1987).
63. Richard Moore, *The Royal Navy and Nuclear Weapons* (London: Frank Cass Publishers, 2001): 39–55.
64. Letter from Coxwell to Deputy Secretary, 10 July 1947, ADM 116/5715, TNA (London).
65. Letter from Deputy Secretary to Coxwell, 11 July 1947, ADM 116/5715, TNA (London).
66. Letter from Admiralty to Barlow (Treasury), Ref. W.G.F.417/47, 5 August 1947, ADM 116/5715, TNA (London).
67. Minute from U.S.F (Morris), 28 October 1947, ADM 116/5715, TNA (London).
68. Ibid.
69. Very little archival material of Brundrett's survives, although some of his papers exist at the Churchill archive centre. For a general synopsis of his career, see "Profile: Sir Frederick Brundrett," *New Scientist* (9 January 1958): 16–17. He is also extensively discussed in Robert Clayton and Joan Algar, ed., *A Scientist's War: the war diary of Sir Clifford Paterson, 1939–1945* (London: Peder Peregrinus Ltd., 1991).
70. Memorandum, Deacon for Brundrett, 7 February 1948, ADM 116/5715, TNA (London).
71. Ibid.
72. Memorandum, Deacon for Brundrett, 6 October 1947, ADM 116/5715, TNA (London).
73. Memorandum, Deacon for Brundrett, 7 February 1948, ADM 116/5715, TNA (London).
74. It is difficult to ascertain Wyatt's true intentions, as he left no written record of his feelings in a memoir or correspondence with others. All that remains is meeting minutes.
75. Minutes of a meeting, 'Reconstitution of the Discovery Committee and Establishment of National Institute of Oceanography', 11 February 1948, ADM 116/5715, TNA (London).
76. Memo by J.N. Carruthers 16 February 1948, ADM 116/5715, TNA (London).
77. Letter from CRNSS to Head of M. Br., 3 March 1948, ADM 116/5715, TNA (London).
78. Ibid.

79. Minute for D.C. (R&D), 14 October 1947, ADM 116/5715, TNA (London).
80. Memo by Mackintosh sent to Jourdain, 20 Feburary 1948, received in Cabinet office 1 March 1948, CAB 124/555, TNA (London).
81. Note by Head of Military Branch (Abercrombie), 13 July 1948, ADM 116/5715, TNA (London).
82. Grove, *Vanguard to Trident,* 39.
83. Sean M. Maloney, *Securing Command of the Sea: NATO Naval Planning 1948–1954* (Annapolis, MD: Naval Institute Press, 1995): 135.
84. Minute signed by Wyatt, 29 November 1945, ADM 116/5715, TNA (London).
85. Report of the Sub-Committee for Oceanography, Nov. 1944, CAB 124/555, TNA (London); for a history of the Liverpool Tidal Institute, see Anna Carlson-Hyslop, "An Anatomy of Storm Surge Science at Liverpool Tidal Institute 1919–1959: Forecasting, Practices of Calculation and Patronage" (PhD diss., University of Manchester, 2010).
86. Minute, 25 January 1946, BJ 5/276, TNA (London).
87. The executive committee consisted of Deacon, Brundrett and other "allies". At this time it would be wrong to say that this was an officially established body; it was certainly taking advantage of the NOC's embryonic state.
88. Margaret Deacon, "Founding of the NIO," *Of Seas and Ships,* 32.
89. Phrase adapted from Peder Roberts, "A Frozen Field of Dreams: Science, Strategy, and the Antarctic in Norway, Sweden and the British Empire, 1912–1952" (PhD diss., Princeton, 2010): 129.

References

Agar, Jon, Brian Balmer. "British Scientists and the Cold War: The Defence Research Policy Committee and Information Networks, 1947–1963." *Historical Studies in the Physical and Biological Sciences* 28 (1998): 209–252.

Bennett, Brett M., and Joseph M. Hodge, eds. *Science and Empire: Knowledge and Networks of Science Across the British Empire, 1800–1970.* London: Palgrave Macmillan, 2011.

Calvert, Jane. "What's Special About Basic Research?" *Science, Technology and Human Values* 31 (2006): 199–220.

Carlsson-Hyslop, Anna. "An Anatomy of Storm Surge Science at Liverpool Tidal Institute 1919–1959: Forecasting, Practices of Calculation and Patronage." PhD diss., University of Manchester, 2010.

Clarke, Sabine. "Pure Science with a Practical Aim: The Meanings of Fundamental Research in Britain, Circa 1916–1950." *Isis* 101 (2010): 285–311.

Clayton, Robert, Joan Algar, ed. *A Scientist's War: The War Diary of Sir Clifford Paterson, 1939–1945*. London: Peder Peregrinus Ltd., 1991.

Deacon, George E.R. "Physical and Chemical Problems in Oceanography." *Proceedings of the Linnaean Society of London* 158 (1947): 80–82.

Deacon, Margaret. "Steps Toward the Founding of the NIO." In *Of Seas and Ships and Scientists: The Remarkable Story of the UK's National Institute of Oceanography*. Edited by Anthony Laughton, John Gould, M.J. 'Tom' Tucker, Howard Roe. Cambridge: Lutterworth Press, 2010: 25–26.

Edelsten, H. "John Claud Fortescue Fryer. 1886–1948." *Obituary Notices of Fellows of the Royal Society* 7 (1950): 94–106.

Edgerton, David. *Britain's War Machine: Weapons, Resources, and Experts in the Second World War*. Oxford: Oxford University Press, 2011.

Endersby, Jim. *Imperial Nature: Joseph Hooker and the Practices of Victorian Science*. Chicago: University of Chicago Press, 2008.

Feldberg, W.S. "Henry Hallet Dale, 1875–1968." *Biographical Memoirs of Fellows of the Royal Society* 16 (1970): 77–174.

Grove, Eric. *Vanguard to Trident: British Naval Policy Since Second World War*. Annapolis, MD: Naval Institute Press, 1987.

Gummett, Phillip J., Geoffrey L. Price. "An Approach to the Central Planning of British Science: The Formation of the Advisory Council on Scientific Policy." *Minerva* 15 (1977): 119–143.

Hardy, Alistair C. "Obituary. Stanley Wells Kemp F.R.S. 1882–1945." *Journal of the Marine Biological Association of the United Kingdom* 26 (1946): 219–234.

Hughes, Jeff. "Doing Diaries: David Martin, the Royal Society and Scientific London, 1947–1950." *Notes & Records of the Royal Society* 66 (2012): 273–294.

Jones, R.V., W.S. Farren. "Henry Thomas Tizard, 1885–1959." *Biographical Memoirs of Fellows of the Royal Society* 7 (1961): 313–348.

Katz, B. "Archibald Vivian Hill. 26 September 1886–3 June 1977." *Biographical Memoirs of Fellows of the Royal Society* 24 (1978): 71–149.

Kline, Ronald. "Constructing 'Technology' as 'Applied Science': Public Rhetoric of Scientists and Engineers in the United States, 1880–1945." *Isis* 86 (1995): 194–221.

MacKenzie, John M. *Propaganda and Empire*. Manchester: Manchester University Press, 1986.

Maloney, Sean M. *Securing Command of the Sea: NATO Naval Planning 1948–1954*. Annapolis, MD: Naval Institute Press, 1995.

Maurice, H.G. "Discussion on Oceanography." *Proceedings of the Linnaean Society of London* 158 (1947): 78–79.

Moore, Richard. *The Royal Navy and Nuclear Weapons*. London: Frank Cass Publishers, 2001.

New Scientist. 'Profile: Sir Frederick Brundrett.' 9 January 1958.

Newitt, D.N. "Alfred Charles Glyn Egerton. 1886–1959." *Biographical Memoirs of Fellows of the Royal Society* 6 (1960): 39–64.

Pantin, C.F.A. "Robert Beresford Seymour Sewell, 1880–1964." *Biographical Memoirs of Fellows of the Royal Society* 11 (1965): 146–155.

Piekle, Roger, Jr. "'Basic Research' as a Political Symbol." *Minerva* 50 (2012): 339–361.

Piller, G.J. "Tizard, Sir Henry Thomas (1885–1959)." *Oxford Dictionary of National Biography*. Oxford University Press, 2004. doi:https://doi.org/10.1093/ref:odnb/36528 [accessed 14/12/14].

Ratcliffe, J.A. "Edward Victor Appleton, 1892–1965." *Biographical Memoirs of Fellows of the Royal Society* 12 (1966): 1–21.

Roberts, Peder. "A Frozen Field of Dreams: Science, Strategy, and the Antarctic in Norway, Sweden and the British Empire, 1912–1952." PhD diss., Stanford University, 2010.

Roberts, Peder. *The European Antarctic: Science and Strategy in Scandinavia and the British Empire*. New York: Palgrave Macmillan, 2011.

Schauz, Désirée. "What Is Basic Research? Insights from Historical Semantics." *Minerva* 52 (2014): 273–328.

CHAPTER 4

Collaboration on Defence, Intelligence, and Internationalism During the 1950s

Our object in the future should be to get international agreement to survey the world for the good of the world. (Report of the Royal Society Sub-Committee for Oceanography, 24 August 1944[1])

The idea of transforming oceanography into a global science through international collaboration appeared a pressing issue for British oceanographers in the 1950s. Transnational oceanographic studies were tentatively enacted through the International Geophysical Year (IGY) and the establishment of other new international oceanographic bodies. The IGY comprised a series of international studies focusing on the earth and its environments, "collaboratively" produced in 1957–58 by nations from both sides of the Iron Curtain. These collaborative projects attempted to produce a synoptic view of Earth phenomena that was unrestricted by national borders or ideologies. Its most famous outcome was the launching of the Sputnik satellite in 1957, which took the scope of the research programme beyond Earth. One of the key figures in this internationalisation movement leading to the organisation of the IGY was George Deacon.

This imagining of ocean science embodied the shift from oceanography of, and for, the nation towards a science that was, rhetorically, *sans frontières*. International collaborations allowed Deacon to strengthen his post-war connections at home and rise to the role of principle representative of British oceanography internationally. The consequences of

S. A. Robinson, *Ocean Science and the British Cold War State*, Palgrave Studies in the History of Science and Technology, https://doi.org/10.1007/978-3-319-73096-7_4

this situation were twofold. First, international collaboration within oceanography helped Deacon to distance the NIO from the traditional naval research, testing, and development that the Admiralty research establishments instructed the new institute to carry out. Although Deacon complied with the Admiralty's requirements he believed that these studies would not help the NIO to elaborate innovative research. He thus used international collaboration as a way to retain control of the NIO's research agenda. Second, he used his role as an international broker in oceanographic studies in the service of other departments in the British government. These circumstances made Deacon a contributor to intelligence-gathering operations concerning advances in Soviet oceanography. In essence by the end of the 1950s Deacon was the "hybrid" historical actor at the centre of the British military oceanography network. In this capacity, he was not only a scientist and negotiator within British science, but also an international diplomat and intelligence gatherer for the British state.

The NIO at the Admiralty Research Laboratories (ARL)

The NIO was founded, on paper at least, in 1949, but it took until 1954 for the Institute to move into its own facilities at Wormley, Surrey.[2] Prior to this it was housed within the Admiralty Research Laboratories at Teddington, south-west London.[3] These initial five years were difficult for the fledgling institute. The British economy was still struggling to recover from the war and rationing remained in force.[4] The Royal Navy was no longer the "senior service" for Britain's strategic defence, a role relinquished to the nuclear bomber-equipped RAF. During the Korean War (1950–53) the Navy had acted in a supporting role in a conflict which had demonstrated that there remained a requirement to maintain a modern navy, but not at its pre-war size.[5] One consequence of the Korean War and the return of Churchill to power (re-elected October 1951) was a move to restart research and development activities. However, budgets were tight and the testing and calibrating of new Admiralty equipment replaced the fundamental scientific research that Deacon wished to prioritise.[6]

The emphasis on testing marked an important divergence in the ways in which Deacon and senior officers at the Admiralty Research Laboratory thought about the NIO's functions.[7] Yet, Deacon knew that an open conflict with these senior officers could be detrimental to the NIO. While the physical oceanographers continued to develop and strengthen their network within the Admiralty, biological oceanographers challenged their colleagues' newly acquired influence and sought to check their emerging power. This was a period in which Deacon had to solidify his position as director of the Institute by cultivating the personal links which collaboration both inside and outside the military brought.

Research collaboration with the Navy was undertaken against the backdrop of general scepticism amongst some members of the Admiralty as to the need for civilian oceanographers at all.[8] Justifying the existence of the NIO during its early years was contingent on persuading Admiralty officials and Royal Navy officers, scientists, and technicians, that studies of the ocean environment were essential for developing offensive and defensive weapons technologies.[9] The physical oceanography programme of the early 1950s was essentially a continuation of the W group's work on waves, currents, and confluence of ocean water masses.[10] Nevertheless with limited test facilities of its own, the NIO staff were often seconded onto other projects rather than having the resources to develop their own ideas.[11] Although elsewhere other collaborations, such as that between the Department of Geodesy and Geophysics at the University of Cambridge and the Hydrographic Office which encompassed the Challenger Expedition (1950–52), were largely successful and did demonstrate that hydrographic and oceanographic research could work hand in hand (Fig. 4.1).[12]

Despite the challenges Deacon was mostly satisfied with this initial arrangement, although he voiced concerns to the Chief of the Royal Navy Scientific Service that the Admiralty did not seem to grasp the full potential of basic oceanographic research.[13] There were several benefits to this early arrangement: new staff gained experience of working with the military at somebody else's expense, and their time was chargeable to the different research establishments leading the projects.[14] As part of one of these secondments the newly recruited Henry Charnock was sent to the Mediterranean to work on a project studying the detection of submarine wake.[15] Charnock had trained as a meteorologist with the RAF during the Second World War, working on marine weather and air–sea interaction, and this had prompted him to study for a degree in meteorology after the war at Imperial College before entering the NIO as a marine meteorologist in 1949.

Charnock was employed in the role of "expert", working as part of a larger team, which included engineers, naval officers, and experimental

Fig. 4.1 NIO Physical Oceanographers at the ARL c.1950. (*left to right*) **Back:** Norman Smith, Frank Pierce, Cyril Williams, Rick Hubbard, D.W. 'Dick' Privett, Laurence Baxter, Leon Verra. **Front:** Jim Crease, M.J. 'Tom' Tucker, Henry Charnock, George Deacon, Ken Bowden, Jack Darbyshire. (Image from the Archives of the National Oceanographic Library, National Oceanography Centre. Southampton.)

staff from elsewhere. The success or failure of the project was not Deacon's primary concern. It gave Charnock the opportunity to network with colleagues in other branches of the Royal Navy Scientific Service, and if the project was successful then some credit would be won for the contribution of the NIO staff; however if the project turned out to be a failure Deacon could distance himself and his staff, claiming that they were merely consultants.[16] Once these collaborations were an established part of the NIO programme they could be used to advocate for increases in resources, staff, and financial support.[17]

The greatest challenge was the lack of a permanent home for the NIO. Ideally a potential site for the Institute would facilitate the development of an independent research programme and geographical separation from the existing Admiralty research machinery. Paradoxically the best way to lobby for separation was to demonstrate how indispensable the NIO was to Admiralty research needs. In 1952 Deacon wrote a memorandum arguing for the NIO's 'need for its own home,' in doing so making extensive use of Charnock's defence work on submarine wake detection as an example of the value of the Institute.[18] During these experiments basic oceanographic measurements had led to the detection of thermal changes in the water disturbed by the wakes of submarines at both the surface and periscope depth. Even though this was only a preliminary trial, Deacon was keen to point out that the collaboration between the Royal Navy, the ARL, and members of the Institute had been successful even if it had put the Institute's own research 'out of gear to achieve it' (thus suggesting that the Admiralty's needs would always be put first). [19]

But Deacon was also unhappy about how the collaboration impacted on the staff workload. In his memorandum he carefully pointed out that the Institute's own research programme using the *Discovery II* would be much improved if men like Charnock had not been required for a whole year. The need to argue for a degree of separation challenged the notion of the NIO as a useful scientific-defence institute for the Admiralty that was on tap, rather than fully independent. Deacon wanted more staff so that he could maintain the defence side of the Institute whilst simultaneously developing an independent research programme. This would ensure that good scientists who wished to pursue their own research projects would be attracted to the NIO.[20]

Increasing involvement in defence research resulted in Deacon being invited to join committees within the Admiralty. Whilst his more junior scientists were involved in trials and projects directed from elsewhere within the Admiralty, from early 1950 Deacon was invited to sit on various panels overseeing research, assessing projects, and other research group work within the Navy.[21] This allowed Deacon into the heart of its research community and made the NIO director a part of, rather than separate from, the management of naval R&D as a whole. Most importantly it gave Deacon equal status with the other directors of Admiralty research establishments, sections, and laboratories. Whilst these new positions gave Deacon some influence within the Admiralty research policy community, it also made him more dependent upon decisions taken centrally in the Admiralty and restricted his ability to elaborate a programme of his own.

Being inside the research management machinery, Deacon came to know the priorities and the pressure to focus on defined research priorities. It also brought him into close proximity to the Directorate of Physical Research (DPR). The DPR, established in 1946, was responsible for Admiralty scientific research into mathematical and physical problems. Accompanied by Charnock, Deacon began to forge a close working relationship with the DPR as the body whose work most closely mirrored that of oceanographers. The DPR was essentially a very similar organisation to the NIO in terms of its purpose; the Admiralty could have called it the Directorate of Military Oceanography, but this would have been seen as overly narrowing its focus. The DPR's work was primarily focused on developing technologies that would harness the oceans as a physical environment in order to improve the capabilities of Royal Navy weapons systems.[22] This would be achieved via the management of basic research applied to technological outcomes. In the early 1950s this predominantly involved trials of new anti-submarine weapons technologies and exploring new ways of detecting submarine movements. Much of this work used the same methodologies and tools as the oceanographers.

The first project that the NIO and the DPR worked on together was a trial of an anti-submarine radar developed by the Marconi Company.[23] The DPR approached Deacon because they needed to understand the prevailing sea states at the test site. This required measurements of sea wave heights, experiments the W group had been carrying out since 1944.[24] This project came at a useful juncture as it allowed new members of the Institute to gain experience of using the wave recorders developed by Deacon and others, at the DPR's expense.[25] More significantly it brought the NIO into the Admiralty machine, prompting the assistant to the Director of Physical Research to write to Deacon thanking him for his staff's invaluable assistance and cooperation.[26] Deacon was again approached when a year later it was proposed that there should be joint Anglo-Canadian studies into infra-red detection of a submarine's wake.[27] This relied on measuring changes in the thermal properties of the water displaced by the movement of a submarine. From the very first meeting (3 March 1952) Deacon and Charnock were invited to attend alongside colleagues from the DPR and the Canadian Defence Research Board.[28]

Building on this new collaboration, Charnock sent a circular to the DPR on 11 March 1952 arguing that the Admiralty should assist in an existing collaboration the NIO had with the Meteorological Office to take bathythermographic observations from weather ships. This was being hindered because of lack of adequate instrumentation.[29] Charnock's suggestion was for the DPR to lend some of their instruments for experiments that would help both departments. In reality this was a resurrection and modernisation of a 1930s ICES project that had been led by James Carruthers in the North Sea using British, German, and Dutch light ships to take current measurements.[30] In this context it represented a practical way in which the DPR could repay the work of the NIO in other areas—the relationship was becoming reciprocal.

The cordiality between military scientists and civilian experts did not last long. In December 1952 Deacon agreed with the DPR that the following summer trials would be conducted in the Mediterranean studying the possibility of thermal wake detection of submarines.[31] In return for services rendered the DPR would pay the NIO £2,300 and Deacon agreed to develop a special shallow-depth bathythermograph for a further £1,500.[32] This was a lucrative arrangement. Not only was the NIO getting extra money for doing precisely what the Admiralty paid them for already, but they would be working in the Mediterranean with the full support of the Royal Navy which would not tie up the institution's existing ships. Studies into the thermal layering of the oceans were becoming an NIO research speciality, enabling further development of the NIO's own research agenda.

However, unlike the previous Marconi trials the new trials did not proceed smoothly. The arrangements proved not as simple as they had first appeared. When Charnock saw the vessel the Royal Navy was offering to conduct the trials in, he realised that it was completely unsuited to survey work.[33] Matters quickly deteriorated, threatening the carefully developed collaboration between the NIO and the existing Admiralty research structures. Deacon composed an angry letter to William Cook, the Chief of the Royal Navy Scientific Service (CRNSS), who had taken over the position at the end of Brundrett's mandate in 1950,[34] stating that the ship would need a month of modifications before it could be used, which would tie up his staff.[35] Cook had worked on guided weapons before being recruited to the RNSS by Brundrett and was appointed Director of Physical Research in 1947, despite having no background in ocean science. He nonetheless managed to improve the range of ASDIC detection through projects in the Directorate before succeeding his recruiter to become CRNSS.[36] Cook was not a faceless bureaucrat but knew the challenges of technical development and the work of the DPR first hand. Deacon also raised other concerns surrounding the hitherto neglected question of whether the detection equipment actually worked at all. He wrote:

> I find that this project, which this year appears to be mainly designed to find under what oceanographical conditions equipment made by TRE [Telecommunications Research Establishment; Ministry of Supply], which is still in an experimental stage, will detect submarines, is likely to take the most useful effort of three members of our staff for six months. A delay might be useful in several ways. TRE could make sure that their apparatus will in fact detect submarines. I understand that such work is in progress. By next year it may be possible to use the 'Discovery II' and avoid all the uncertainty as to whether an Admiralty vessel and TRE will be ready on September 1st.[37]

Deacon's letter was not well received. Cook stated from the outset: 'your letter has placed me in rather a difficulty'.[38] It transpired that when Deacon had agreed to undertake the work for the DPR on behalf of the NIO the circumstances were already well known; only the exact start date of the trials remained undecided.[39] Additionally TRE were adamant that their equipment did function, backing this up with evidence delivered to Cook.[40] This meant there was no way that a postponement could be justified on the basis that TRE was not ready. Anti-submarine research was quickly becoming the research priority for the ARL and was regarded as an extremely urgent and pressing matter by the defence-minded Prime Minister Winston Churchill.[41]

Cook had no choice but to rein Deacon in, informing him: 'I am sorry about this, Deacon, but it seems to me that we let the Admiralty down if we back out at this stage, having once said we could do it, and the Admiralty have gone ahead on this basis'.[42] Deacon was forced to reluctantly concede 'my outcry was rather hasty; it is now clear that everyone is rallying round, and Pierce and Charnock are flying to Malta to-morrow to make a start'.[43] The result of his outcry was that Deacon was forced to deploy his staff three weeks early to ensure they could meet the start date. However, the NIO was not the only group to try and withdraw from the scheme, and others were to be more successful.

The Services Electronics Research Laboratory (SERL) provided the key link between the scientists doing the testing and research (NIO), and the technicians undertaking the development at TRE. Writing to Charnock on 29 July, SERL withdrew its earlier offer of support in terms of manpower—but suggested that there might be a way of loaning instruments to the NIO scientists. Confidently they proclaimed that if the NIO wished to take issue with this they could always 'take the matter up at a high level'.[44] SERL's decision to back out was not surprising. SERL's chief scientist Peter Wright wrote in his controversial book *Spycatcher* that it was at this time when working alongside the Marconi Company—whose equipment the NIO was first brought in to test back in early 1952—that he and SERL began drifting towards investigation into the workings of newly developed spy "bugs" that the US Central Intelligence Agency (CIA) had asked Marconi to study.[45] For Wright this led to a complete career change when in 1954 he left SERL and its anti-submarine work to take up the post of Chief Scientist at the British home intelligence service MI5.[46]

With no way out Deacon reluctantly sent his scientists out to Malta to undertake trials of somebody else's untested equipment, without the support of other independent technical staff.[47] The result was a total failure; TRE's infra-red system, dubbed the "yellow duckling" was, actually, a white elephant. When Smith of the DPR wrote to Deacon to thank his staff for their competency and efficient day-to-day organisation of the trials but also to ask for Charnock's attendance at a final informal meeting to discuss the future of the device, Deacon was adamant that his institution's involvement had ended.[48] The NIO had never been the principal agency involved in the project yet at the time they were the scapegoat for its failure.

This was not the end of the matter. In a letter to the Director of Physical Research at the Admiralty, Deacon complained bitterly about focus on developing the technology first and understanding the ocean second.[49] He did not miss the opportunity to re-affirm the centrality of oceanographic work in the activities of the network he had sought to establish. He concluded his heated letter: 'one can only argue that all governments are spending quite a lot of money on projects under the headings in my first paragraph (anti-submarine warfare, harbour defence, submarine warfare, navigation, costal engineering, and fisheries) and they will save money and find greater value from what they spend if some of it is used to study the medium which can no longer be regarded as a perfect fluid.'[50]

Criticising his opposite number in the Admiralty research machine reflected a shift in the position of the NIO. In 1954 Deacon held an open day at Wormley to formally open the NIO's laboratories there; writing to Cook (CRNSS), he boasted of eighty visitors. He had achieved the oceanographers' primary goal—a fully-fledged institute funded by the Royal Navy but very much separate from it.[51] By 1954 Deacon had further strengthened his relationships abroad, especially in the USA, and viewed the Admiralty's approach to NIO's research programmes as straightjacketing. He therefore sought to use his international connections to secure greater independence in his management of oceanography at home, and the forthcoming International Geophysical Year offered the perfect opportunity to do just that.

The National in the Transnational: The NIO and the International Geophysical Year

The International Geophysical Year (IGY) was a programme of international scientific projects that ran from 1 July 1957 until 31 December 1958. Seeking to promote the earth sciences through a global synoptic picture of the Earth, it involved both Eastern and Western bloc countries, at the same time publicly subverting Cold War rhetoric through the universality of science above political squabbles. The presentation of the IGY as a scientific utopia devoid of politics was very important for its organisers. Lloyd Berkner, the Vice-President of the Special Committee for the International Geophysical Year, famously stated that:

> The plan has been organized in a severely divided world, but politics has not hindered the needed collaboration. If occasionally politics has reared its head a little, it has been dismissed with a shrug and a smile by men of all nations who knew that the growth of knowledge need not be hampered by partisan views.[52]

This claim about the absence of politics has been the launch point for early historical studies of the IGY.[53] In reality, however, the project was directly linked to politics and the patronage of science.[54] Reliance on finance, organisation, and resources linked science and military organisations together, imposing national military and political expectations and accountability on an event that was outwardly presented as "international" and above such national concerns.

The build-up to the IGY allowed George Deacon to take a more active role in international scientific diplomacy as secretary of the IGY subcommittee for oceanography. As Jacob Hamblin has argued, the IGY gave several oceanographers the opportunity to forge careers more akin to that of the scientific diplomat serving on several overlapping committees and bodies whilst advocating at the supra-national level for further international collaboration and cooperation.[55] Deacon used these international opportunities to better his position within the British scientific community, the government machine, and the national security infrastructures. The IGY changed Deacon's career trajectory from national actor with a key role establishing oceanography at a national level but with limited international links from his military work (primarily with US, Canadian and Norwegian oceanographers) to an internationally recognised oceanographic leader with new contacts, particularly in Asia and behind the Iron Curtain.

The IGY represented the culmination of ten years' rebuilding of the links between scientists which had been severed during the Second World War alongside a general reconstruction of science financed in the West by the United States.[56] Independent of US efforts, the British had attempted to reconnect with their European counterparts.[57] In the early 1950s Deacon relied on colleagues in the Institute, such as James Carruthers, to maintain or establish international links with European military and civilian oceanographers.[58] In 1950 Carruthers returned to Bergen in Norway, and started to work on an inclined current meter for studies of fisheries hydrography—a key first step in re-establishing links with Norwegian oceanographers and foreign oceanographic instrument makers such as Bergen-Nortek.[59] In 1952, Carruthers travelled to the Netherlands to meet scientists who had taken part in pre-war studies on the sea bottom, particularly sedimentary analysis.[60] This was important for mine warfare, and the initial impetus for his travel came from the Hydrographer and the chief of the Royal Navy Scientific Service, rather than George Deacon. He wrote a detailed report of his trip on which he acquired equipment that could be copied and experimented with by NIO staff.[61] While Deacon had better links with US oceanographers, Carruthers' European links from his fisheries work made him uniquely qualified to connect the new institute into his existing network of North Sea oceanographic nations.[62]

The IGY promised to go beyond traditional bilateral collaboration to something unprecedented in ocean science: transnational collaboration. While it did not mark a radical departure from ongoing naval research, it allowed resources to be acquired that would not otherwise have been available. The IGY was first discussed in the USA in 1950 during a dinner organised by Lloyd Berkner for the visit of British ionospheric scientist Sidney Chapman.[63] As a consequence of the IGY, Deacon's network of acquaintances expanded, with connections to many more scientists, officials, and military-intelligence operatives within and beyond Britain. It would be all too easy to use Hamblin's argument about the IGY expanding international careers as a departure point here, leaving the national context to gravitate towards a trans- or even supra-national context.[64] However, the career of a scientific diplomat like Deacon was built primarily upon the support of others, especially those in government who provided patronage on the national stage, enabling opportunities in international science.

The initial support for a British oceanographic programme for the IGY came from two sources: the Royal Society (which represented Britain on the International Council of Scientific Unions, ICSU); and Admiral Archibald Day, the new Hydrographer of the Royal Navy, who was having reservations about the direction of ocean research in the service. Day felt that there was too much focus on developing technologies at the expense of basic research into the oceans themselves, an attitude very much in tune with Deacon's.[65] Day encouraged Deacon to travel to the second IGY meeting in Rome in 1954 as the British observer.[66] Wider interest in the IGY programme from within the Admiralty only arose following the publication of the American programme.

Returning from the Rome meeting Deacon wrote to Day to excitedly inform him that 'I found that the US oceanographers intend after all to support the International Geophysical Year in a big way'.[67] In August 1954 a formal copy of the American proposals for the IGY was received, indicating the non-scientific reasoning for undertaking such a massive project:

> The United States has accepted, for better or for worse, world-wide responsibility in social, economic, and political matters. To meet those responsibilities it must look on the oceans as a medium that connects North America with other continents, and it must seek an understanding of what the resources of the ocean can mean to mankind.[68]

With firm American commitment, Western European nations now began to formulate their own programmes in time for the third IGY meeting in Brussels in 1955.

This was also the first time that Soviet oceanographic plans for the IGY were revealed. They proposed to use thirteen vessels, which made up 19 per cent of the entire worldwide IGY oceanographic fleet, the largest being the *Vityaz*, at 6000 tons far larger than any Western vessel.[69] Rather than contribute to the study of ocean currents or further the concept of dynamical oceanography being developed by Western physical oceanographers, the Soviets intended to undertake a systematic exploration of ocean trenches. To Deacon this was mere "hydrography", a more detailed study of deep places that the British *Challenger* expedition had already started in the 1870s. More recently, in 1951, HMS *Challenger* had recorded the deep record of 5,899 fathoms.[70]

However, the Russians were looking for deep stagnant water, which was exactly what the Anglo-American expedition was trying to disprove the existence of. In this way, they hoped to identify possible sites for the dumping of nuclear waste.[71] For the Soviets, therefore, national needs associated with the expansion of their nuclear energy programme were the instigating factor for their collaboration with the IGY. This objective was given greater scientific legitimacy by being combined with a biological question of whether marine organisms existed in the deepest parts of the ocean given the extreme pressures. Ultimately this part of the operation proved unsuccessful when the difficulties of performing a trawl eight miles deep from a surface vessel being buffeted by waves frustrated Soviet efforts.[72] But the scope of these plans and the tools that the Soviets deployed caused Western militaries to panic. Analysing Soviet capabilities but being unable to work closely with Soviet scientists was a major challenge for Western intelligence services. The IGY was an opportunity not only to further national programmes using the resonance of international collaboration, but also to learn more about the progress of foreign oceanography programmes.

As individual national programmes were circulated to leading oceanographers in other countries, Deacon and others in a similar position were eager to take stock of the interests of other nations in oceanography. It was the synoptic nature of their studies that would bring them together—a term derived from meteorology where synoptic weather charts had been produced since the mid-nineteenth century. The "buzzword" for the IGY, it meant that any data produced by any nation during the period of study would allow for the production of a unified vision of the earth—a measured snapshot in time which highlighted countries with similar research objectives, shared areas of interest, and innovative instruments. It was this knowledge of foreign science that was to prove of so much interest to the intelligence community.[73]

In 1954 Deacon had been appointed secretary of a committee elected to formulate oceanographic plans for the IGY, and when oceanography was subsequently adopted as part of the IGY this committee became a working group coordinating ocean science projects. Through his correspondence with oceanographers reporting their scientific activities, Deacon expanded his international network of contacts. For example, prior to the IGY the British oceanographic community had had few dealings with Japanese oceanographers, aside from producing a report on Japan's oceanographic programme during the war, but during the IGY Deacon became one of the first Western oceanographers to discover that the Americans had managed to test a hydrogen bomb close to the Japanese oceanographic fleet, causing their radiation sensors to surpass the level at which it was safe to continue operating. These sensors had been put in place following the 1954 *Lucky Dragon* incident when fallout from a US nuclear test had irradiated Japanese fishermen in the Pacific.[74]

Despite these setbacks the IGY oceanography programme was particularly successful for the NIO, and ultimately strengthened UK–US oceanographic links. The main NIO contribution to the IGY programme concerned the dynamism of the oceans, and attempted to settle a long-running debate about whether the oceans were truly stagnant at great depths as had often been assumed.[75] Two pre-war surveys had suggested that this assumption was false. The German *Meteor* expedition had discovered that the composition of seawater varied greatly with depth, suggesting that different bodies—or layers—of water sat one upon another.[76] Henry Stommel of WHOI had been studying the theoretical question of how, if the Gulf Stream, an easily identifiable current in the North Atlantic, moved such a great volume of water, there had to be an opposite and equal movement of water so that the oceans remained in equilibrium.[77] He put forward the idea of equal and balanced currents, driven by differences in water temperature and salinity.[78]

The second pre-war survey was that carried out by George Deacon. As part of the Discovery Committee's investigations between 1929 and 1939, Deacon had identified a sub-Antarctic upwelling of seawater.[79] Stommel's model suggested an opposite effect with sub-Arctic water. However, taking measurements of seawater at depth was difficult, because the instrument had to stay at a prescribed depth and then be allowed to drift whilst being tracked from the surface. If a shallow instrument drifted in one direction, and another deeper one in the opposite direction, Stommel's theory of deep currents and ocean circulation would be proven. To solve this problem John Swallow, one of Deacon's newest appointments at the NIO, developed the neutral-buoyance float (see Fig. 4.2).[80] The device was designed to sink to a predefined depth and then drift on any current that existed. A battery-powered sonic "pinger" relayed the

Fig. 4.2 The first neutrally buoyant float for tracking water movements at depth was developed by John Swallow, a British oceanographer. It consisted of an aluminium pipe with a battery and timer circuit that would excite a magnetostrictive transducer, a "pinger", hanging underneath. John Swallow pictured, 1955. (Image from the Archives of the National Oceanographic Library, National Oceanography Centre, Southampton.)

position of the float to the surface, allowing the speed and direction of the drift to be recorded.

This was first attempted in March 1957, when several floats were deployed off the *Discovery II* set to sink to various depths, then tracked by the ship, whose position could be verified using Loran stations on shore (a primitive electronic navigation system accurate to about a quarter of a mile).[81] This initial experiment found that floats deployed at 6,500 feet were almost stationary, whilst those at 9,000 feet moved south at 8 miles per day, showing that there was a deep current present and that a shallower current mixed with the deeper current at around 6,500 feet, cancelling each other out. However the question remained as to whether the seawater moved at the depth of the sea floor. To solve this problem

Deacon's other new recruit Anthony Laughton developed a device to sit on the sea floor and relay the movements of a caged ball back to the surface using cameras—very different work from his doctoral work on marine sediments.[82] This instrument demonstrated that even at a depth of 10,500 feet the current was still moving at 5 miles per day. These experiments were carried out in conjunction with WHOI using their vessel *Atlantis*.[83]

These joint Anglo-American expeditions demonstrated that even the biggest questions that remained unanswered about the oceans could be answered, given sufficient resources. However, mirroring concerns he would raise later about space research, Deacon felt that the launching of space satellites overshadowed this work, which had answered a significant unknown question about the dynamic of the ocean circulation. Walter Sullivan, the IGY chronicler, quoted Deacon as having likened the work of the oceanographers to the launching of satellites which were destined to orbit the earth for years. Deacon is said to have remarked that the Swallow floats would do much the same thing in that although their pingers had fallen silent they would continue to float around the ocean at their predetermined depths for years, just as dead satellites would orbit the earth.[84] For Swallow a series of publications followed and he quickly gained an international reputation for his float, which was good for the Institute as the sales of instruments brought in an income, and the ocean circulation programme was distinct from work being carried out anywhere else in the UK. Most importantly this research suggested that thermoclines existed in deeper water, and that various phenomena that existed near the surface also existed below, further complicating the hydrography that would affect the installation of fixed seafloor surveillance equipment (see Chap. 5).[85] For Deacon the IGY signalled the success of the NIO, his institution's Gulf Stream project being the largest collaboration British oceanographers had made during the year, but it also gave him a sense of perspective about oceanography globally. This knowledge was to prove useful in non-scientific contexts.

Whilst this work was producing "intelligence" from the oceans, the IGY also supplied open-source intelligence about foreign oceanographic capabilities and equipment. The unprecedented access to and information about foreign science that the IGY enabled Deacon to gather was extremely useful to the intelligence community within Britain. These events and Deacon's central contribution to them on the international stage placed him in a remarkably good position to know about Soviet advances in oceanography and thus enhance his position as an intelligence gatherer.

Spying on Soviet Oceanography

In 1958 the Royal Navy distributed a confidential document entitled "Oceanography and defence in the USSR, 1956–1958".[86] This document was the result of scientific intelligence gathering, or more simply *spying* on rival science projects of the Russian oceanographers. It provided one of the earliest intelligence summaries of the capabilities of Soviet ocean science, including details about the size of the Soviet oceanographic fleet, in particular its large vessels, and the fact that only they possessed a deep-water magnetic studies vessel, other nations such as Britain having abandoned construction of such ships.[87] This report was not the product of espionage as such (as this phraseology might suggest), rather it was the product of what intelligence analysts call open-source intelligence (or OSINT).[88] Today this type of data collection is now more commonly associated with mining the Internet using vast supercomputers, whereas its 1950s equivalent was collecting data from unclassified sources, scientific literature, and meetings.[89] So useful and important was this method that the British had a whole body devoted to this type of intelligence gathering: the Joint Intelligence Bureau (JIB). Although a JIB officer at the Admiralty authored the aforementioned report, it was only made possible by the IGY and George Deacon's participation in the organisation of the oceanography programme. That the British security services were interested in gathering intelligence on Soviet oceanography demonstrates the growing significance of oceanographic knowledge in the organisation of naval defence, surveillance, and anti-submarine warfare operations.

During the early Cold War the British developed methods for collecting information about Soviet science and in doing so created a vast network of informants from within the British scientific community. In 1947, following a review of intelligence gathering during the Second World War, the Joint Intelligence Committee noted that 'the less money we have to spend on preparations for war, the more important it is to have a first-class intelligence service in peacetime'.[90] Furthermore, if Britain was not to be surprised by the scientific advances of other nations it would have to be in a position to better calibrate its scientific investment, making economical use of its scientific resources.[91] Following the report the JIC established the JIB with the objective of using 'the existing machinery for obtaining secret intelligence and [to]...obtain through channels of its own, where existing channels are inappropriate "overt" intelligence below the grade secret'.[92] Effectively this tasked the JIB with gathering OSINT for the purposes of national security. As Huw Dylan has demonstrated, the creation of the JIB highlighted that in the case of scientific intelligence, 'secrecy did not necessarily equate with importance'.[93] This orientated the JIB towards gathering knowledge rather than secrets. From 1947 until restructuring in 1964, the JIB worked towards the construction of a national knowledge bank of scientific, economic, and topographical intelligence that was of defence relevance to military commanders and policy-makers concerned with directing operations and policy.[94] Turning scientists into "informants" or, to put it less provocatively, suppliers of OSINT about their foreign scientific colleagues, challenged the scientific ideal of the independence of science from politics.[95]

The spreading of international collaboration in science, given the Cold War need for increased intelligence and vigilance, has often been seen as a challenge with restrictions imposed upon science and scientists by regimes of classification and security.[96] The classification of scientific data and restrictions placed on the free movement of scientific knowledge across frontiers were part of an attempt to deprive the Soviets of free access to Western science. As Doel and Needell write, 'there are fundamental differences between the ideals and values of intelligence gathering and the ideals and values that serve to define the international community of scientists'.[97]

Yet these very contradictions between openness and secrecy, internationalism and national security, often had to co-exist even in the same 'laboratory quadrangles'. This meant that international collaborative projects represented an opportunity for intelligence gathering and the scientists involved in them could play a role—controversially—as gatherers of intelligence.[98] The British Director of Scientific Intelligence, R.V. Jones, wrote, 'the reluctance of reputable scientists to give up active work in science for a spell in intelligence is understandable. Some may well feel that it is a degrading activity for their talents, and that it is at best a dirty business, prying into other countries' secrets.'[99] Several historians, including Ron Doel and Peder Roberts, have recently demonstrated that during the Cold War senior scientists could be employed as intelligence agents in specific circumstances.[100]

George Deacon played a similar role throughout the IGY, though his work with the intelligence community pre-dated the IGY. Reports that the Soviets had presented research on Pacific oceanography at a meeting in Copenhagen in 1953 prompted George Turney of the Division of Scientific Intelligence (DSI) of the Ministry of Defence to begin asking

British oceanographers to assess the 'extent of their [Soviet] effort'.[101] This initial request for expert analysis of enemy capabilities came not directly to Deacon but through the Cambridge geophysicist Edward Bullard. But Bullard was not as well placed as Deacon to know about plans for oceanography work in the Soviet Union, as he was still the Director of the National Physical Laboratory. Deacon's new role in the IGY meant that he could gain that information far more easily; the fruitful relationship he established with Turney is documented by their extensive correspondence.[102] Deacon played down his role as intelligence gatherer, claiming that his knowledge of Soviet efforts in the field of oceanography was based solely on rumours and there was little that the NIO had been able to find within the available published sources that was particularly ground breaking, scientifically.[103] Nevertheless Turney was invited to come down to the NIO regularly in the late 1950s, and meet Deacon to learn more about the IGY activities.[104]

The correspondence between Deacon and Turney, albeit at the suggestion of Bullard, demonstrates the continuing influence of wartime networks within the Admiralty scientific community during the Cold War. Turney knew Bullard as they had both worked at the Royal Navy Mine Design Department, HMS *Vernon,* during the war.[105] Turney was an expert in magnetometer research and the degaussing of ships.[106] In 1945 Turney met with Deacon as they both worked under the Royal Navy Scientific Service. Deacon was the senior scientist at the ARL leading the W group, and Turney was the Assistant Director of the service directly under Charles Seymour Wright. During the late 1940s Turney moved away from scientific research towards scientific intelligence via the MoD Division of Scientific Intelligence, where he chaired the committee overseeing research into unidentified flying objects (UFOs).[107] From 1952, when R.V. Jones was appointed by Churchill to lead the DSI, interests within the division diversified under Jones' "mantra" that 'scientific Intelligence, with its constant vigil for new applications of science to warfare by the enemy, [is] the first watchdog of national defence'.[108] Under Jones' leadership the DSI came to see the collection of scientific intelligence as contributing to analysis under the simple equation quoted in Goodman: 'THREAT = CAPABILITY + INTENT'.[109] Whilst intelligence specialists could predict "intent", academic/scientific "experts" were relied upon to assist with the understandings of capability, which is how Deacon and the NIO became useful to the intelligence community.

Published scientific papers continued to be one of the key sources of intelligence on Soviet oceanography. Following initial meetings in 1954 Turney began to provide the NIO with copies of Soviet and Eastern bloc scientific papers on ocean science that were translated by Foreign Office personnel and made available at no cost to a number of intelligence departments and the NIO library.[110] Meanwhile Deacon's position as secretary of the IGY oceanography committee helped to produce "synoptic" intelligence by putting together the translated literature and the new information provided to Deacon by the Soviet oceanographers at scientific meetings. By 1956 others such as Michael Longuet-Higgins, who had joined the Institute in 1954, were also involved in reading and interpreting Soviet capabilities through the study of scientific literature.[111] Deacon remained pessimistic about the capabilities of Russian scientists, stating again that he had 'not found anything as advanced as work in the USA or UK'.[112] The fact that the unknown author of the confidential Royal Navy report into Soviet oceanography was also dismissive of these capabilities suggests that his pamphlet was informed by the secret activities of Turney and Deacon.[113]

Deacon's disparaging view of the capabilities of Soviet oceanography has been documented by Jacob Hamblin.[114] As Hamblin has argued more recently, this had perhaps more to do with the publication methods of the Soviet oceanographers, who sailed with printing presses on board and rushed to print their scientific data as they steamed back to port.[115] Seeing oceanography as the measurement of the oceans rather than as an object of which oceanographers should ask questions often irked Western oceanographers, Deacon in particular.[116] Assessing Soviet science from its outputs was only one way to gather intelligence on the capabilities of enemy oceanography. In the build-up to the IGY not only did Soviet scientists tour British facilities but they also provided detailed plans for their participation in the event which was a good indicator of research capacity regardless of capability.

Under the auspices of the international ideals of science, Soviet scientists began to tour foreign countries (with KGB agents following them at a suitably short distance) to gain intelligence on Western science. The appearance of Soviet scientists at British establishments caused renewed concern for Turney, but also showed just how deeply intelligence was embedded within oceanographic work. Concern about two Soviet oceanographers who had been working at Cambridge, Treshnikov and Maximov, was communicated via an American scientist, Russell Raitt of the University of California.[117] Raitt spent a year at Cambridge, and promptly relayed what he knew to Turney and (via Turney) to Deacon.[118] These two scientists had also visited the NIO. Deacon informed Turney that, 'Treshnikov was obviously not a scientist, and did not seem to follow what was going on. Maximov was very bright and was particularly interested in our harmonic analyser and the new "floats" which we can set to drift at predetermined depths.'[119] Of course, concern about Soviet spies was omnipresent in British science following the Cambridge spy-ring scandal that had broken at the beginning of the decade. The NIO had been implicated as the so-called sixth member of the ring, Alistair Watson, had been sent, in lieu of more severe punishment, to work out the rest of his career at Wormley at the behest of MI5.[120]

It was not just the visits of Soviet scientists to Britain that enabled OSINT to be gathered. Participation in international scientific collaboration, whilst fostering and strengthening Western bloc cohesion, also provided an opportunity to interact, network, and gather information on non-Western science. Deacon's privileged position as the secretary of the IGY working group on oceanography allowed him to meet and interact with foreign oceanographers, and he was keen to use this opportunity both for Britain's intelligence agencies but also for his own ambitions. He felt the Russians were 'using the IGY oceanography programme to show what a lot of science they are doing'.[121] He wrote 'if one has the money ocean exploration is easy, relatively useful, and a good way of showing the flag, and showing how keen you are on science'. The potential benefit that the IGY offered in this case to Deacon was to extend his national network. Linking intelligence operatives to this network was just another layer of justification for continued funding. Deacon accordingly continued writing to Turney about Soviet IGY plans, and especially their wish to survey sea areas which had strategic significance in the Cold War conflict:

> I enclose two notes on the IGY plans, but I don't think they will help you to judge whether the Russians have an ulterior motive in going near California. It is obviously important for them to know as much as possible about the density layering etc. if they want to send submarines there, but our kind of research will in the end be more useful than exploration which has reached the stage of diminishing returns.[122]

Despite Deacon's reluctance to voice opinions on Soviet intentions and motivations, his correspondence with Turney continued throughout the

IGY. It was the intelligence that Deacon was able to share as the secretary of the working group that was particularly useful to the DSI. As secretary he was privy to information about plans, surveys, and instruments in advance of their publication in academic journals. The collection of documents detailing the IGY plans of other nations was an excellent source for an intelligence official.

Whilst Deacon was able to share this material with Turney, in return the NIO was provided with "free" translations of Russian papers, and in its turn the DSI then received an assessment of the content and potential of Soviet scientific advances.[123] Turney offered Deacon more than just scientific literature, paving the way—when needed—for Deacon to find a firmer place in the corridors of power by supporting his views on several occasions when—as we shall see in the next chapters—he was increasingly being challenged within the military–science policy network that he had helped to establish.

The 1958 DSI report on Soviet oceanography was careful to disguise the sources and informants whose collection and assessment activities it was based on. Actual human intelligence gathering was only referred to twice. The report acknowledged that one ocean scientist whose open publications they had relied heavily upon, Vasiliy Vladimirovich Shuleykin, had disappeared since it had been announced that he had been relieved of his post as director of the Marine Hydrophysics Institute in Moscow. When Western oceanographers had visited the institute they had asked about his whereabouts and relayed the replies that he was either ill or at sea back to the DSI. The conclusion of the report also betrayed the role of British oceanographers in the analysis of Soviet science: 'Western oceanographers, who have studied the Russian publications and have had personal discussions with their leading workers, regard the results disclosed as in no way commensurate with the effort in men and ships which has been expended.'[124] The extent to which Deacon contributed to the writing of the DSI report is difficult to assess, but the correspondence between Deacon and Turney does make it clear that the NIO director was involved in the assessment of Shuleykin's work. He provided comments on papers translated at the Foreign Office and assessed Soviet work on waves.[125]

The gathering of scientific intelligence on Soviet oceanography was intended for a much larger audience than the British intelligence community. The 1958 DSI report was distributed to non-military government departments such as the Foreign Office and MI5, along with similar intelligence organisations amongst the Dominions (Australia, Canada, and

New Zealand) but also to the CIA.[126] Supplying the United States with intelligence went beyond this report alone. Early in 1958 Turney once again wrote to Deacon asking for assistance in gathering information about ocean science in the Eastern bloc. Responding to a request from the USA for information on an agreement between the USSR, China, North Korea, and Vietnam regarding cooperation in fisheries, oceanography, and limnology in the Western Pacific, Deacon was asked to supply any information that Turney could 'pass on to our friends'.[127] Deacon's reply is unfortunately contained within an as yet still classified file held by the National Oceanography Centre library archive.[128]

The gathering, analysis, and publication of reports relating to scientific intelligence on Soviet ocean science continued after the IGY. In February 1960 a second report was published by the JIB, "A Survey of Recent Soviet Underwater Acoustic and Oceanographic Studies to the Beginning of 1960".[129] Based on forty Soviet published papers this second report was heavily focused on the military aspects of underwater acoustics research, but also discussed plans to construct an oceanographic research submarine to explore the underwater world. Once again the sources and scientists who provided the analysis for the DSI were carefully concealed. The methodology was the same, the report stating 'this survey of acoustic and oceanographic studies are based entirely upon the writer's summaries, reported herein, of literature published openly by Soviet investigators'.[130] The inclusion of the NIO in the distribution of this report strongly indicated that the institute once again provided some of the summaries upon which the report was constructed and interpreted by intelligence specialists.

Compromise and International Routes Away from National Challenges

International opportunities offered by large-scale collaborations such as the IGY enabled Deacon to break away from national research dictated by other government research groups. Initially, though perhaps naïvely, Deacon was more than willing to take an active part in Navy trials, which seemed to present clear advantages. Other establishments were willing to pay for NIO expertise, boosting the Institute's income and justifying increasing staffing. Moreover, the trials gave relatively inexperienced oceanographers sea time at somebody else's expense. Finally, sea trials offered the opportunity to acquire new advanced instrumentation at somebody else's expense and, even more usefully, to test equipment without having to pay for it or retain the instruments if they proved obsolete. However, as issues around NIO involvement later demonstrated that these expeditions could prove political minefields, Deacon sought to plan research that could be based at the NIO, but would also be fully under his control and not imposed upon him by other naval research departments in Britain.

In shifting the research agenda away from the national towards larger multi-national projects, Deacon was able to play national and international commitments off against one another, arguing that he could not commit resources and manpower to certain national projects that had become burdensome, or pushed the NIO away from doing ocean science towards work more traditionally undertaken by Admiralty research laboratories. He was also inclined, when needed, to avoid international commitments that he did not strongly support, advocating national pressures to prioritise research in the national arena.[131] As Hamblin has argued, his "diplomacy" role marked a fundamental shift in the focus of his career away from organising the day-to-day activities of the NIO's programme towards a role that saw him working to enable others to get oceanography "done". We shall explore this in more detail over the following chapters.

International oceanography came of age in the 1950s. In this period, ocean science was carried out with the full backing of the military, particularly in terms of resources and finance, which marked a departure from existing relations and opened new avenues of intelligence and surveillance. These new opportunities strengthened the position of the NIO at the heart of the state bringing the institution and especially its director into contact with a new network of actors who in turn offered political support, translated papers, and access to government or military resources. Perhaps most importantly Deacon gained evidence of the usefulness of oceanographers to the Cold War state. Over the following chapters this shift is further explored, charting both the advantages and disadvantages experienced by members of the NIO in the deepening of Cold War relationships with the state-military machinery.

Notes

1. Report of the Sub-Committee for Oceanography, 24 August 1944, CAB 124/555, TNA (London).
2. Details of the debate surrounding the location of the institute are provided in Chap. 3.
3. "Profile: Dr G.E.R. Deacon – The Study of Movement in the Seas", *New Scientist* (15 September 1960): 717.

4. Till Geiger, *Britain and the Economic Problem of the Cold War: The Political Economy and the Economic Impact of the British Defence Effort, 1945–55* (Farnham: Ashgate, 2004).
5. Stephen Prince, "The Contribution of the Royal Navy to the United Nations Forces during the Korean War," *Journal of Strategic Studies*, 17:2 (1994): 96.
6. Here I am using the term fundamental research in the way in which Deacon would have understood it, namely pure science with an applied aim. See Sabine Clarke, "Pure Science with a Practical Aim: The Meanings of Fundamental Research in Britain, circa 1916–1950," *ISIS*, 101 (2010): 285–311.
7. Jon Agar and Brian Balmer, "British Scientists and the Cold War: The Defence Research Policy Committee and Information Networks, 1947–1963," *Historical Studies in the Physical and Biological Sciences,* 28:2 (1998): 236–7.
8. Letter from Deacon to Buckingham (Director of Physical Research, Admiralty), 22 January 1954, GERD Papers D4/1, NOC Library (Southampton); Letter from Admiral Day (Hydrographer) to Deacon, 10 September 1954, GERD Papers, D5/1, NOC Library (Southampton).
9. Memo by George Deacon, 'National Institute of Oceanography – Need for its own home', 1 August 1952, GERD Papers, D1/1, NOC Library (Southampton).
10. George E.R. Deacon, "The growth and decay of waves," *Assoc. Océanogr. Phys, Procès-Verbaux* (1952): 82–3; Jack Darbyshire, "The generation of waves by wind," *Proc. Roy. Soc. A.,* (1952): 299–328; Jim Crease, "The origin of ocean currents," *J. Inst. Navig.,* (1952): 280–4.
11. Letter to Deacon from Smith (Department of Physical Research, Admiralty), 10 February 1950, GERD Papers, D5/1, NOC Library (Southampton); Letter from C.S. Durst (Meteorological Office of the Air Ministry), 3 July 1951, GERD Papers, D5/1, NOC Library (Southampton).
12. There is insufficient space to go into a full account of the Challenger Expedition here. However it is worth looking at George Stephen Ritchie, *Challenger: The Life of a Survey Ship,* (New York: Abelard-Schuman, 1958), and George Stephen Ritchie, *No Day Too Long – An Hydrographer's Tale* (London: Hyperion Books, 1989).
13. Letter Deacon to Cook (CRNSS), 3 March 1952, *Secret,* GERD Papers, D1/3, NOC Library (Southampton).
14. Letter from Smith (Department of Physical Research, Admiralty) to Deacon, 15 December 1952, GERD Papers, D5/1, NOC Library (Southampton). The NIO was to be reimbursed £2,300 to 'cover salaries, subsistence, travelling etc.'

15. Letter from Deacon to Smith (Department of Physical Research, Admiralty), 22 December 1952, GERD Papers, D5/1, NOC Library (Southampton).
16. Letter from Deacon to Smith (Department of Physical Research, Admiralty), 1 February 1954, GERD Papers, D5/1, NOC Library (Southampton).
17. Deacon's first priority was to find a new home for the Institute away from Teddington, where biologists and physicists could work together; see memo by Deacon 'National Institute of Oceanography – need for its own home', January 1952, GERD Papers, D1/1, NOC Library (Southampton).
18. Memo by Deacon 'National Institute of Oceanography – need for its own home', January 1952, GERD Papers, D1/1, NOC Library (Southampton).
19. 'NIO Oceanographical Defence Research Work', memo by Deacon 'National Institute of Oceanography – need for its own home', January 1952, GERD Papers, D1/1, NOC Library (Southampton).
20. Ibid.
21. Letter from Smith (Department of Physical Research) to Deacon, *Secret*, GERD Papers, D5/1, NOC Library (Southampton). Deacon joined the Scientific Advisory Group of the Admiralty Detection Establishment (essentially the centre of ASW R&D); records of these meetings are also contained within GERD, D5/1.
22. Similar bodies existed in the US Navy; see Gary E. Weir, *An Ocean in Common: American Naval Officers, Scientists, and the Ocean Environment* (College Station TX: Texas A&M University Press, 2001).
23. Letter from DC Fakley (on behalf of the Director of Physical Research, Admiralty) to Deacon, 4 January 1952, GERD Papers, D5/1, NOC Library (Southampton).
24. See Norman F. Barber, "The Behaviour of Waves on Tidal Streams," *Proceedings of the Royal Society, A*, 198 (1949); George E.R. Deacon, "Waves and Swell," *Quarterly Journal of the Royal Meteorological Society*, 75:325 (1949): 227–38; Norman F. Barber, "Ocean Waves and Swell," *Institute of Civil Engineers* (1950).
25. Letter from DC Fakley (on behalf of the Director of Physical Research, Admiralty) to Deacon, 4 January 1952, GERD Papers, D5/1, NOC Library (Southampton).
26. Ibid., 'The assistance of the NIO is invaluable and their willing cooperation is very much appreciated.'
27. Notes of meeting held on 3 March 1952, GERD Papers, D5/1, NOC Library (Southampton). Present at this meeting were four Canadians,

three members of the DPR, Deacon and Charnock (NIO), and four members of the Ministry of Supply.

28. Ibid.
29. Circular for the DPR by Henry Charnock, 11 March 1952, GERD Papers, D5/1, NOC Library (Southampton).
30. L.R. Lumby, J.N. Carruthers, "An Experiment dealing with the Interpretation of Salinity Distribution Charts," *J. Cons. Int. Explor. Mer.*, (1928): 176–90; J.N. Carruthers, A.L. Lawford, V.F.C. Veley, "Continuous Observations from Anchored Vessels on Water Movements in the Open Sea," *Deutsche Hydrographische Zeitschrift* 3:5/6 (1950): 276–88; J.N. Carruthers, H.J. Garrood, "A new Current Measuring Instrument for the Purposes of Fishery Research," *J. Cons. Int. Explor. Mer.*, (1926): 127–39; J.N. Carruthers, "The Water Movements in the Straits of Dover. Exceptional Currents in Winter 1929–30," *J. Cons. Int. Explor. Mer.* (1930): 167–191.
31. Reference Sheet from Directorate of Physical Research (H.Smith – Scientist) to Deacon, 'Proposed Thermal Wake Detection Trails', 15 December 1952, GERD Papers, D5/1, NOC Library (Southampton).
32. Ibid.
33. Letter from Deacon to Cook (CRNSS), 1 July 1953, GERD Papers, D5/1, NOC Library (Southampton).
34. Brundrett became Deputy Chief Scientific Adviser to the Ministry of Defence in 1950, serving under Tizard and Cockcroft before becoming Chief Scientific Adviser himself, a post he held from 1954 to 1960, when replaced by Sir Solly Zuckerman.
35. Letter from Deacon to Cook (CRNSS), 1 July 1953, GERD Papers, D5/1, NOC Library (Southampton).
36. "Profile: Sir William Cook", *New Scientist* (3 December 1959): 1126–1127.
37. Letter from Deacon to Cook (CRNSS), 1 July 1953, GERD Papers, D5/1, NOC Library (Southampton).
38. Letter from Cook (CRNSS) to Deacon, 6 July 1953, GERD Papers, D5/1, NOC Library (Southampton).
39. Letter from Cook (CRNSS) to Deacon, 6 July 1953, GERD Papers, D5/1, NOC Library (Southampton).
40. Letter from Cook (CRNSS) to Deacon, 6 July 1953, GERD Papers, D5/1, NOC Library (Southampton): 'TRE have stated that their equipment is functioning correctly and we have seen supporting evidence.'
41. Annual reports of the Admiralty research laboratory http://arl.g3w1.com/Corsair/refMat.htm; John W. Young, *Churchill's Last Campaign: Britain and the Cold War, 1951–5* (Oxford: Clarendon Press, 1996); John W. Young, *The Foreign Policy of Churchill's Peacetime Administration,*

1951–55 (Leicester: Leicester University Press, 1988); Ritiche Ovendale, *British Defence Policy Since 1945* (Manchester: Manchester University Press, 1994): 97; Christopher M. Bell, *Churchill and Sea Power* (Oxford: Oxford University Press, 2013): chapter 11.

42. Letter from Cook to Deacon, 6 July 1953, GERD Papers, D5/1, NOC Library (Southampton).
43. Letter from Deacon to Cook, 10 July 1953, GERD Papers, D5/1, NOC Library (Southampton).
44. Letter from Ross (SERL) to Charnock, 29 July 1953, GERD Papers, D5/1, NOC Library (Southampton).
45. Peter Wright, *Spycatcher* (New York: Viking Penguin, 1987): 18–23.
46. Wright, *Spycatcher,* 27.
47. Letter from Deacon to Cook (CRNSS), 10 July 1953, GERD Papers, D5/1, NOC Library (Southampton).
48. Letter from H. Smith (DPR) to Deacon, 14 January 1954; Letter from Deacon to H.Smith, 1 February 1954, GERD Papers, D5/1, NOC Library (Southampton).
49. Letter from Deacon to Buckingham (Director of DPR), 22 January 1954, GERD Papers, D4/1, NOC Library (Southampton).
50. Ibid.
51. Letter from Deacon to Cook (CRNSS), 14 July 1954, GERD Papers, D4/1, NOC Library (Southampton).
52. Lloyd V. Berkner, "The International Geophysical Year, 1957–58: A Pattern for International Cooperation in Research," *Proceedings of the American Philosophical Society,* 101:2 (1957): 159–163.
53. Such as Walter Sullivan, *Assault on the Unknown: The International Geophysical Year* (New York: McGraw-Hill, 1961).
54. Ibid.
55. Hamblin, *Oceanographers and the Cold War,* 99.
56. The post-war reconstruction of European science on the back of US finance has been well documented by John Krige, *American Hegemony and the Postwar Reconstruction of Science in Europe,* (MIT Press, 2006). On the origins and purpose of the IGY for the rebuilding of international science, see Fae L. Korsmo, "The Genesis of the International Geophysical Year," *Physics Today,* 60 (July 2007); Gregory A. Good, "Sydney Chapman: Dynamo behind the International Geophysical Year," in *Globalizing Polar Science: Reconsidering the International Geophysical Years,* ed. Roger D. Launius, James R. Fleming, David H. DeVorkin (New York: Palgrave Macmillan, 2010): 177–204; Allan A. Needell, "Lloyd Berkner and the International Geophysical Year Proposal in Context: With some Comments on the Implications for the Comité Spéciale de l'Année Géophysique Internationale, CSAGI, Request for

Launching Earth Orbiting Satellites," in *Globalizing Polar Science: Reconsidering the International Geophysical Years,* ed. Roger D. Launius, James R. Fleming, David H. DeVorkin (New York: Palgrave Macmillan, 2010): 205–224; Dian O. Belanger, "The International Geophysical Year in Antarctica: A Triumpth of 'Apolitical' Science, Politics and Peace," in *Globalizing Polar Science: Reconsidering the International Geophysical Years,* ed. Roger D. Launius, James R. Fleming, David H. DeVorkin (New York: Palgrave Macmillan, 2010): 265–278.

57. Initially this had been carried out in pursuit of Nazi science and technology which began even before the cessation of hostilities in 1945; see Sean Longden, *T-Force: The Race for Nazi War Secrets, 1945* (London: Constable & Robinson, 2009); Jones, R.V., *Most Secret War: British Scientific Intelligence 1939–45* (London: Hodder & Stoughton, 1979).
58. Carruthers not only went to Germany at the end of the war, but was also communicating with Dutch oceanographers during the early 1950s. See J.N. Carruthers, 'Memorandum for Hydrographic Department', 17 March 1952, GERD Papers, D5/1, NOC Library (Southampton).
59. James N. Carruthers, "On the Instrumental Measurement of Line Shape under Water," *Deutsche Hydrographische Zeitschrift* (1954): 22–35; James N. Carruthers, "Some Inter-relationships of Oceanography and Fisheries," *Archiv Fur Meteorologie, Geophysik und Biokilmatologie, Serie B,* 6:1–2 (1954):167–89.
60. "Memorandum for Hydrographic Department by JN Carruthers, 'Mine Hunting: sedimentary aspects: report on a visit to the Netherlands 9th to 12th March 1952'," *Secret,* GERD Papers, D5/1, NOC Library (Southampton).
61. Ibid.
62. Through his pre-war fixed drift current measurements in the North Sea under ICES.
63. Fae L. Korsmo, "The Birth of the International Geophysical Year," *The Leading Edge,* 26 (2007): 1312–1316.
64. Jacob D. Hamblin, "'The Navy's 'Sophisticated' Pursuit of Science: Undersea Warfare, the Limits of Internationalism, and the Utility of Basic Research, 1945–1956," *Isis,* 93:1 (2002): 1–27; Jacob D. Hamblin, "Gods and Devils in the Details: Marine Pollution, Radioactive Waste, and an Environmental Regime circa 1972," *Diplomatic History,* 32:4 (2008): 539–560.
65. Letter from Vice Adml Sir Archibald Day (Hydrographer of the Royal Navy) to Deacon, 10 September 1954, GERD Papers, D5/1, NOC Library (Southampton).

66. Letter from Vice Adml Sir Archibald Day (Hydrographer of the Royal Navy) to Deacon, 18 May 1954, GERD Papers, D5/1, NOC Library (Southampton).
67. Letter from Deacon to Vice Adml Sir Archibald Day (Hydrographer of the Royal Navy), 30 June 1954, GERD Papers, D5/1, NOC Library (Southampton).
68. 1st American Programme, received 30 August 1954, GERD Papers, D12/1, NOC Library (Southampton).
69. Walter Sullivan, *Assault on the Unknown*, 362.
70. Hamblin, *Oceanographers and the Cold War*, 82; George Steven Ritchie, *Challenger: The Life of a Survey Ship*, (London: Hollis & Carter, 1957) pp.225–232.
71. Jacob D. Hamblin, *Poison in the Well: Radioactive Waste in the Oceans at the Dawn of the Nuclear Age*, (New Brunswick, NJ: Rutgers University Press, 2008): 129.
72. Walter Sullivan, *Assault on the Unknown*, 362–363.
73. On big data and the IGY, see Elena Aronova, Karen S. Baker, Naomi Oreskes, "Big Science and Big Data in Biology: From the International Geophysical Year through the International Biological Program to the Long Term Ecological Research Program, 1957-present," *Historical Studies in the Natural Sciences*, 40:2 (2012): 183–224
74. Aya Homei, "The Contentious Death of Mr Kuoyama: Science as Politics in the 1954 Lucky Dragon Incident," *Japan Forum*, 25:2 (2013): 212–232; Aya Homei, "Fallout from Bikini: the Explosion of Japanese Medicine," *Endeavour*, 31:4 (2007): 129–133; Toshihiro Hiuchi, "Tipping the Scale of Justice: the Fallout Suit of 1958 and the Environmental Legal Dimension of Nuclear Pacifism," *Peace & Change*, 38:1 (2013): 33–55.
75. George E.R. Deacon, "Oceanography in the International Geophysical Year," *IUGG News Letter*, 5:16 (1956): 541–548.
76. Sabine Höhler, "Depth Records and Ocean Volumes: Ocean Profiling by Sound Technology, 1850–1930," *History and Technology*, 18:2 (2002): 119–154; Eric L. Mills, "Socializing Solenoids: The Acceptance of Dynamic Oceanography in Germany around the Time of the 'Meteor' Expedition," *Historisch-Meereskundlisches Jahrbuch*, (1998): 11–26; Cornelia Lüdecke, "Diverging Currents: Depicting Southern Ocean Currents in the Early Twentieth Century," in *Extremes: Oceanography's Adventures at the Poles*, ed. Keith Benson and Helen Rozwadowski (Sagamore Beach, MA: Science History Publications, 2007): 71–105.
77. Henry Stommel, "A Survey of Ocean Current Theory," *Deep Sea Research*, 4 (1957): 149–184.

78. Henry Stommel, "The Abyssal Circulation," *Deep Sea Research,* 5 (1958): 80–82.
79. George E.R Deacon, *The Hydrology of the Southern Ocean* (Cambridge: Cambridge University Press, 1937). See also Eric L. Mills, "Creating a Global Ocean Conveyer: George Deacon and *The Hydrology of the Southern Ocean,*" in *Extremes: Oceanography's Adventures at the Poles,* ed. Keith Benson and Helen Rozwadowski (Sagamore Beach, MA: Science History Publications, 2007):107–132.
80. John C. Swallow, "A neutral-buoyancy Float for Measuring Deep Currents," *Deep-Sea Research,* 3 (1955): 74–81; John C. Swallow, "Some Further Deep Current Measurements using Neutrally Buoyant Floats," *Deep-Sea Research,* 4 (1957): 93–104; John C. Swallow, B. V. Hamon, "Some Measurements of Deep Currents in the Eastern North Atlantic," *Deep Sea Research,* 6 (1959): 155–168.
81. Walter Sullivan, *Assault on the Unknown,* 353
82. John C. Swallow, L.V. Worthington, "An Observation of a Deep Countercurrent in the Western North Atlantic," *Deep-Sea Research,* 8 (1961): 1–19. This was also used by Japanese oceanographers after the IGY; see Saski, T., Watanabe, S., Oshiba, G., "New Current Meters for Great Depths," *Deep-Sea Research,* 12:6 (1965): 815–824.
83. Walter Sullivan, *Assault on the Unknown,* 351–355.
84. Walter Sullivan, *Assault on the Unknown,* 360.
85. Robinson,A., Stommel, H., "The Oceanic Thermocline and the associated Thermohaline Circulation," *Tellus,* 11 (1959): 295–308.
86. 'Oceanography and Defence in the USSR, 1956–58', [DSI Report No. 143] (1959), DEFE 44/27, TNA (London)
87. The notion of spying on science has been mostly clearly identified by Peter Maddrell, *Spying on Science: Western Intelligence in Divided Germany 1945–61*, (Oxford: Oxford University Press, 2006). At the end of the Second World War the British had also produced a document on Japanese oceanography; HM Stationery Office, 'Oceanography in Japan', British Intelligence Objectives Sub-Committee, (1946) [Manchester University Library – Special Collections]. The British abandoned the construction of their magnetic research vessel, HMS *Research,* after the outbreak of the war; the largest collection of documents regarding the construction of this vessel are held by the National Maritime Museum, Greenwich.
88. Mercado, S.C., "Sailing the Sea of OSINT in the Information Age," *Studies in Intelligence,* 48:3; [https://www.cia.gov/library/center-for-the-study-of-intelligence/csi-publications/csi-studies/studies/vol48no3/article05.html accessed 18/7/14].

89. Hew Dylan, "The Joint Intelligence Bureau: (Not so) Secret Intelligence for the Post-War World," *Intelligence and National Security*, 27:1 (2012): 27–45.
90. Dylan, "The Joint Intelligence Bureau", 28.
91. Ibid., 28.
92. 'DEFE 69/526, CAB 81/131, TNA (London); quoted in Dylan, *The Joint Intelligence Bureau.*
93. Dylan, "The Joint Intelligence Bureau", 33.
94. Ibid., 45.
95. This also happened in the case of sea-ice forecasting where the Scott Polar Research Institute in Cambridge alongside the Foreign Office organised a series of exchanges in the Soviet Union to gather scientific intelligence; see Peder Roberts, "Scientists and Sea Ice under Surveillance in the Early Cold War", in *The Surveillance Imperative: Geosciences during the Cold War and Beyond*, ed. Simone Turchetti and Peder Roberts (Palgrave, 2014): 133–135.
96. Ron E. Doel, "Does Scientific Intelligence Matter?," *Centaurus*, (2010): 311–22; Western distrust of Soviet oceanography has been discussed by Hamblin, *Oceanographers and the Cold War*, 88–98.
97. Ron E. Doel, Allan A. Needell, "Science, Scientists, and the CIA: Balancing International Ideals, National Needs, and Professional Opportunities," *Intelligence and National Security*, 12:1 (1997): 59.
98. David K. van Keuren, "Cold War Science in Black and White: US Intelligence Gathering and its Scientific Cover at the Naval Research Laboratory, 1948–62," *Social Studies of Science*, 31:2 (2001): 208.
99. Jones, R.V., *Most Secret War: British Scientific Intelligence 1939–45* (Hodder & Stoughton; London, 1979): 661.
100. Ron E. Doel, "Scientists as Policymakers, Advisors, and Intelligence Agents: Linking Contemporary Diplomatic History with the History of Contemporary Science," in Thomas Söderqvist, *The Historiography of Contemporary Science and Technology* (Reading, UK: Harwood Academic Publishers) 215–44; Peder Roberts, "Intelligence and Internationalism: The Cold War Career of Anton Bruun," *Centaurus*, 55:3 (2013): 243–263.
101. Letter from Turney (Naval Intelligence) to Deacon, 6 February 1954, *Secret*, GERD Papers, D5/1, NOC Library (Southampton).
102. It would appear from the archival record and recent publications that in the early years of the JIB they were primarily concerned with Russian nuclear science and technology. See the files contained with the series TNA (London), DEFE 44; and Michael S. Goodman, *Spying on the Nuclear Bear: Anglo-American Intelligence and the Soviet Bomb* (Redwood City, CA: Stanford University Press, 2007).

103. Letter from Deacon to Turney, 10 February 1954, *Secret,* GERD Papers, D5/1, NOC Library (Southampton).
104. Letter from Deacon to Turney, 10 February 1954, *Secret,* GERD Papers, D5/1, NOC Library (Southampton).
105. Kelly, H.W.K., "Historical Introduction to Degaussing, Covering Pre-war Work and Development up to the Start of the Campaign against the German Mine," *Electrical Engineers Part 1,* 93:70 (1946) 430–4; Butterworth, A., "'Development and Use of Magnetic Apparatus for Bomb and Mine Location," *Journal of the Insistution of Electrical Engineers – Part II,* 95:48 (1948) 645–52.
106. Turney, G. L., Cousins, G.E., "A Portable Direct-Reading Magneometer," *Journal of Scientific Instruments* (1938) 360–67; E.P. Harrison, G.L. Turney, H. Rowe, H. Gollop, "The Electrical Properties of High Permeability Wires Carrying Alternating Current," *Proc. R. Soc. Lond. A.* (1936) 451–79; Harrison, E.P., Turney, G.L., Rowe, H., "Electrical Properties of Wires of High Permeability," *Nature,* 135: 3432 (1935) 961.
107. David Clarke, *The UFO files: The Inside Story of Real-life Sightings* (London: The National Archives, 2009): 38.
108. Michael S. Goodman, "Jones's Paradigm: The How, Why, and Wherefore of Scientific Intelligence," *Intelligence and National Security* 24:2 (2009): 236–256, 238; Jones, R.V., "Scientific Intelligence," *Journal of the Royal United Services Institution* (1947): 354.
109. Goodman, "Jones's Paradigm", 243 [capitals in original].
110. Letter from Turney to Deacon, 11 May 1956, GERD Papers D12/4, NOC Library (Southampton); these scientific papers remain at the current NOC library residing in red archive boxes amongst the loose papers collected from various sources, the only suggestion of their previous classified status being the red box and stamps detailing their translation by the Foreign Office.
111. Letter from Turney to Deacon, 11 May 1956, GERD Papers D12/4, NOC Library (Southampton).
112. Letter from Deacon to Turney, 14 May 1956, GERD Papers D12/4, NOC Library (Southampton).
113. 'Oceanography and defence in the USSR, 1956–58', [DSI Report No. 143] (1959), DEFE 44/27, TNA (London); the negative comments are contained in the last paragraph of the report.
114. Hamblin, *Oceanographers and the Cold War,* 88–9.
115. Jacob D. Hamblin, "Seeing the Oceans in the Shadow of Bergen Values," *ISIS* 105:2 (2014): 360.
116. Hamblin, "Shadow of Bergen Values", 352–63.

117. It is likely that this was Aleksei Treshnikov and Igor Maksimov (not Maximov), who later in 1956 led the second Soviet Antarctic expedition but the likely incorrect British spelling of Russian names and the lack of any recorded initials makes positive identification difficult. If I am correct then Treshnikov's disinterest in the NIO's activities makes perfect sense as he was a self-styled explorer who led the on-land portion of the endeavour, whereas Maximov's interest would reflect the fact that Maksimov led the marine component of the Soviet expedition, and Deacon's pre-war activities in these waters was well known. Their activities in Cambridge were probably more a reflection of the city being home to the Scott Polar Research Institute where, as shown by Peder Roberts (Roberts, *Scientists and Sea Ice under Surveillance*, 133–135), there was a keen interest in Soviet research.
118. Letter from Turney to Deacon, 11 May 1956, GERD Papers, D12/4, NOC Library (Southampton).
119. Letter from Deacon to Turney, 14 May 1956, GERD Papers, D12/4, NOC Library (Southampton).
120. Wright, *Spycatcher*, 257.
121. Letter from Deacon to Turney, 14 May 1956, GERD Papers, D12/4, NOC Library (Southampton).
122. Letter from Deacon to Turney, 14 May 1956, GERD Papers, D12/4, NOC Library (Southampton).
123. Letter from Turney to Deacon, 15 August 1956, GERD Papers, D12/4, NOC Library (Southampton).
124. 'Oceanography and defence in the USSR, 1956–58', [DSI Report No. 143] (1959), DEFE 44/27, TNA (London).
125. Letter from Turney to Deacon, 15 August 1956; letter from Deacon to Turney, 17 August 1956, GERD Papers, D12/4, NOC Library (Southampton).
126. See distribution list; 'Oceanography and defence in the USSR, 1956–58', [DSI Report No. 143] (1959), DEFE 44/27, TNA (London).
127. Letter from Turney to Deacon, 21 January 1958, *Secret,* GERD Papers, D12/4, NOC Library (Southampton).
128. This is contained within file (GERD D5/2) of the Deacon Papers. A request was made in August 2013 but as yet this file has not been cleared by the MoD, although it has previously been open to researchers and items from this file were referenced in Hamblin (2005).
129. 'A Survey of Recent Soviet Underwater Acoustic and Oceanographic Studies to the Beginning of 1960', [DSI Report No. 170], February 1960, DEFE 44/43, TNA (London).

130. 'A Survey of Recent Soviet Underwater Acoustic and Oceanographic Studies to the Beginning of 1960', [DSI Report No. 170], February 1960, 8, DEFE 44/43, TNA (London).
131. See Chap. 4 and Hamblin, *Oceanographers and the Cold War*, 182.

References

Agar, Jon, Brian Balmer. "British Scientists and the Cold War: The Defence Research Policy Committee and Information Networks, 1947–1963." *Historical Studies in the Physical and Biological Sciences* 28 (1998): 236–7.

Aronova, Elena, Karen S. Baker, Naomi Oreskes. "Big Science and Big Data in Biology: From the International Geophysical Year Through the International Biological Program to the Long Term Ecological Research Program, 1957-Present." *Historical Studies in the Natural Sciences* 40 (2012): 183–224.

Barber, Norman F. "The Behaviour of Waves on Tidal Streams." *Proceedings of the Royal Society, A* 198 (1949): 81–93.

Barber, Norman F. "Ocean Waves and Swell." *Lecture Published by the Institute of Civil Engineers* (1950).

Belanger, Dian O. "The International Geophysical Year in Antarctica: A Triumph of "Apolitical" Science, Politics and Peace." In *Globalizing Polar Science: Reconsidering the International Geophysical Years.* Edited by Roger D. Launius, James R. Fleming, David H. DeVorkin. New York: Palgrave Macmillan, 2010.

Bell, Christopher M. *Churchill and Sea Power.* Oxford: Oxford University Press, 2013.

Berkner, Lloyd V. "The International Geophysical Year, 1957–58: A Pattern for International Cooperation in Research." *Proceedings of the American Philosophical Society* 101 (1957): 159–163.

Butterworth, A. "Development and Use of Magnetic Apparatus for Bomb and Mine Location." *Journal of the Institution of Electrical Engineers – Part II* 95 (1948): 645–52.

Carruthers, James N. "The Water Movements in the Straits of Dover. Exceptional Currents in Winter 1929–30." *Journal du Conseil Permanent International pour l'Exploration de la Mer* (1930): 167–191.

Carruthers, James N. "On the Instrumental Measurement of Line Shape Under Water." *Deutsche Hydrographische Zeitschrift* (1954a): 22–35.

Carruthers, James N. "Some Inter-relationships of Oceanography and Fisheries." *Archiv Fur Meteorologie, Geophysik und Biokilmatologie, Serie B* 6 (1954b): 167–89.

Carruthers, James N., A.L. Lawford, V.F.C. Veley. "Continuous Observations from Anchored Vessels on Water Movements in the Open Sea." *Deutsche Hydrographische Zeitschrift* 3 (1950): 276–88.

Carruthers, James N., H.J. Garrood. "A New Current Measuring Instrument for the Purposes of Fishery Research." *Journal du Conseil Permanent International pour l'Exploration de la Mer* (1926): 127–39.

Clarke, David. *The UFO Files: The Inside Story of Real-life Sightings*. London: The National Archives, 2009.

Clarke, Sabine. "Pure Science with a Practical Aim: The Meanings of Fundamental Research in Britain, Circa 1916–1950." *Isis* 101 (2010): 285–311.

Crease, Jim. "The Origin of Ocean Currents." *Journal of Institute of Navigation* (1952): 280–4.

Darbyshire, Jack. "The Generation of Waves by Wind." *Proceedings Royal Society. A* (1952): 299–328.

Deacon, George E.R. *The Hydrology of the Southern Ocean*. Cambridge: Cambridge University Press, 1937.

Deacon, George E.R. "Waves and Swell." *Quarterly Journal of the Royal Meteorological Society* 75 (1949): 227–38.

Deacon, George E.R. "The Growth and Decay of Waves." *Association d'océanographie physique, Procès-Verbaux* (1952): 82–3.

Deacon, George E.R. "Oceanography in the International Geophysical Year." *IUGG News Letter* 5 (1956): 541–548.

Doel, Ron E. "Scientists as Policymakers, Advisors, and Intelligence Agents: Linking Contemporary Diplomatic History with the History of Contemporary Science." In *The Historiography of Contemporary Science and Technology*. Edited by Thomas Söderqvist. Reading: Harwood Academic Publishers, 1997.

Doel, Ron E. "Does Scientific Intelligence Matter?" *Centaurus* 52 (2010): 311–22.

Doel, Ron E., Allan A. Needell. "Science, Scientists, and the CIA: Balancing International Ideals, National Needs, and Professional Opportunities." *Intelligence and National Security* 12 (1997): 59–81.

Dylan, Huw. "The Joint Intelligence Bureau: (Not so) Secret Intelligence for the Post-War World." *Intelligence and National Security* 27, no. 1 (2012): 27–45.

Geiger, Till. *Britain and the Economic Problem of the Cold War: The Political Economy and the Economic Impact of the British Defence Effort, 1945–55*. Farnham: Ashgate, 2004.

Good, Gregory A. "Sydney Chapman: Dynamo Behind the International Geophysical Year." In *Globalizing Polar Science: Reconsidering the International Geophysical Years*. Edited by Roger D. Launius, James R. Fleming, David H. DeVorkin. New York: Palgrave Macmillan, 2010.

Goodman, Michael S. *Spying on the Nuclear Bear: Anglo-American Intelligence and the Soviet Bomb*. Redwood City, CA: Stanford University Press, 2007.

Goodman, Michael S. "Jones's Paradigm: The How, Why, and Wherefore of Scientific Intelligence." *Intelligence and National Security* 24 (2009): 236–256.

Hamblin, Jacob D. "The Navy's 'Sophisticated' Pursuit of Science: Undersea Warfare, the Limits of Internationalism, and the Utility of Basic Research, 1945–1956." *Isis* 93, no. 1 (2002): 1–27.

Hamblin, Jacob D. *Poison in the Well: Radioactive Waste in the Oceans at the Dawn of the Nuclear Age*. New Brunswick, NJ: Rutgers University Press, 2008a.

Hamblin, Jacob D. "Gods and Devils in the Details: Marine Pollution, Radioactive Waste, and an Environmental Regime Circa 1972." *Diplomatic History* 32 (2008b): 539–560.

Hamblin, Jacob D. "Seeing the Oceans in the Shadow of Bergen Values." *Isis* 105 (2014): 352–363.

Harrison, E.P., G.L. Turney, H. Rowe, H. Gollop. "The Electrical Properties of High Permeability Wires Carrying Alternating Current." *Proceedings of the Royal Society of London, A* (1936): 451–79.

Harrison, E.P., G.L. Turney, H. Rowe. "Electrical Properties of Wires of High Permeability." *Nature* 135 (1935): 961.

Hiuchi, Toshihiro. "Tipping the Scale of Justice: The Fallout Suit of 1958 and the Environmental Legal Dimension of Nuclear Pacifism." *Peace & Change* 38 (2013): 33–55.

Höhler, Sabine. "Depth Records and Ocean Volumes: Ocean Profiling by Sound Technology, 1850–1930." *History and Technology* 18 (2002): 119–154.

Homei, Aya. "Fallout from Bikini: The Explosion of Japanese Medicine." *Endeavour* 31 (2007): 129–133.

Homei, Aya. "The Contentious Death of Mr Kuoyama: Science as Politics in the 1954 Lucky Dragon Incident." *Japan Forum* 25 (2013): 212–232.

Jones, R.V. "Scientific Intelligence." *Journal of the Royal United Services Institution* (1947): 352–360.

Jones, R.V. *Most Secret War: British Scientific Intelligence 1939–45*. London: Hodder & Soughton, 1979.

Kelly, H.W.K. "Historical Introduction to Degaussing, Covering Pre-War Work and Development up to the Start of the Campaign Against the German Mine." *Electrical Engineers Part 1* 93 (1946) 430–4.

Keuren, David K. van. "Cold War Science in Black and White: US Intelligence Gathering and Its Scientific Cover at the Naval Research Laboratory, 1948–62." *Social Studies of Science* 31 (2001): 207–229.

Korsmo, Fae L. "The Birth of the International Geophysical Year." *The Leading Edge* 26 (2007a): 1312–1316.

Korsmo, Fae L. "The Genesis of the International Geophysical Year." *Physics Today* 60 (2007b): 38–43.

Krige, John. *American Hegemony and the Postwar Reconstruction of Science in Europe*. Cambridge, MA: MIT Press, 2006.

Longden, Sean. *T-Force: The Race for Nazi War Secrets, 1945*. London: Constable & Robinson, 2009.

Lüdecke, Cornelia. "Diverging Currents: Depicting Southern Ocean Currents in the Early Twentieth Century." In *Extremes: Oceanography's Adventures at the Poles*. Edited by Keith Benson, Helen Rozwadowski. Sagamore Beach, MA: Science History Publications, 2007.

Lumby, L.R., James N. Carruthers. "An Experiment Dealing with the Interpretation of Salinity Distribution Charts." *Journal du Conseil Permanent International pour l'Exploration de la Mer* (1928): 176–90.

Maddrell, P. *Spying on Science: Western Intelligence in Divided Germany 1945–61*. Oxford: Oxford University Press, 2006.

Mills, Eric L. "Socializing Solenoids: The Acceptance of Dynamic Oceanography in Germany Around the Time of the 'Meteor' Expedition." *Historisch-Meereskundlisches Jahrbuch* 5 (1998): 11–26.

Mills, Eric L. "Creating a Global Ocean Conveyer: George Deacon and *The Hydrology of the Southern Ocean*." In *Extremes: Oceanography's Adventures at the Poles*. Edited by Keith Benson, Helen Rozwadowski. Sagamore Beach, MA: Science History Publications, 2007.

Needell, Allan A. "Lloyd Berkner and the International Geophysical Year Proposal in Context: With some comments on the implications for the Comité Spéciale de l'Année Géophysique Internationale, CSAGI, Request for Launching Earth Orbiting Satellites." In *Globalizing Polar Science: Reconsidering the International Geophysical Years*. Edited by Roger D. Launius, James R. Fleming, David H. DeVorkin. New York: Palgrave Macmillan, 2010.

New Scientist. "Profile: Dr G.E.R. Deacon – The Study of Movement in the Seas." 15 September 1960.

Ovendale, Ritiche. *British Defence Policy Since 1945*. Manchester: Manchester University Press, 1994.

Prince, Stephen. "The Contribution of the Royal Navy to the United Nations Forces During the Korean War." *Journal of Strategic Studies* 17 (1994): 94–120.

Ritchie, George Stephen. *Challenger: The Life of a Survey Ship*. New York: Abelard-Schuman, 1958.

Ritchie, George Stephen. *No Day Too Long – An Hydrographer's Tale*. London: Hyperion Books, 1989.

Roberts, Peder. "Intelligence and Internationalism: The Cold War Career of Anton Bruun." *Centaurus* 55 (2013): 243–263.

Roberts, Peder. "Scientists and Sea Ice Under Surveillance in the Early Cold War." In *The Surveillance Imperative: Geosciences During the Cold War and Beyond*. Edited by Simone Turchetti, Peder Roberts. London: Palgrave, 2014.

Robinson, A., H. Stommel. "The Oceanic Thermocline and the Associated Thermohaline Circulation." *Tellus* 11 (1959): 295–308.

Saski, T., S. Watanabe, G. Oshiba. "New Current Meters for Great Depths." *Deep-Sea Research* 12 (1965): 815–824.

Stommel, Henry. "A Survey of Ocean Current Theory." *Deep Sea Research* 4 (1957): 149–184.

Stommel, Henry. "The Abyssal Circulation." *Deep Sea Research* 5 (1958): 80–82.

Sullivan, Walter. *Assault on the Unknown: The International Geophysical Year.* New York: McGraw-Hill, 1961.

Swallow, John C. "A Neutral-Buoyancy Float for Measuring Deep Currents." *Deep-Sea Research* 3 (1955): 74–81.

Swallow, John C. "Some Further Deep Current Measurements Using Neutrally Buoyant Floats." *Deep-Sea Research* 4 (1957): 93–104.

Swallow, John C., B.V. Hamon. "Some Measurements of Deep Currents in the Eastern North Atlantic." *Deep Sea Research* 6 (1959): 155–168.

Swallow, John C., L.V. Worthington. "An Observation of a Deep Countercurrent in the Western North Atlantic." *Deep-Sea Research* 8 (1961): 1–19.

Turney, G.L., G.E. Cousins. "A Portable Direct-Reading Magnetometer." *Journal of Scientific Instruments* (1938): 360–67.

Weir, Gary E. *An Ocean in Common: American Naval Officers, Scientists, and the Ocean Environment.* College Station, TX: Texas A&M University Press, 2001.

Wright, Peter. *Spycatcher.* New York: Viking Penguin,1987.

Young, John W. *The Foreign Policy of Churchill's Peacetime Administration, 1951–55.* Leicester: Leicester University Press, 1988.

Young, John W. *Churchill's Last Campaign: Britain and the Cold War, 1951–5.* Oxford: Clarendon Press, 1996.

CHAPTER 5

Oceanographers, Surveillance, and Defence Research

> *The origin of this project…emanated from the Americans on the basis of a philosophy of the need for world-wide surveillance with which we ourselves do not agree.* (Admiral Sir J. Peter L. Reid (Third Sea Lord and Controller of the Navy) to Ian Orr-Ewing M.P. (Civil Lord of the Admiralty), 17 August 1961[1]

The pursuit of surveillance drove military strategic agendas throughout the Cold War and was a decisive factor in shaping relations within alliances.[2] Both US and Soviet military agencies strove for integrated monitoring systems that would ensure some level of global coverage as the conflict evolved to include more regions of the world. But this ambition to reach out and monitor globally was not accepted by all of the superpowers' allies. So far as Britain was concerned, this global ambition contrasted greatly with what British military planners sought to achieve through reconnaissance. Their scepticism was partly rooted in the history of British surveillance operations.

Britain's "hidden hand" had been responsible for monitoring abroad for more than a century and represented a pillar of its crumbling empire. During the Second World War, the pursuit of signals intelligence (SIGINT)[3] had extended British intercept networks across the world and as a result the British chiefs of staff had advocated coordinating surveillance operations with US agencies, and specifically in strategically vital

S. A. Robinson, *Ocean Science and the British Cold War State*,
Palgrave Studies in the History of Science and Technology,
https://doi.org/10.1007/978-3-319-73096-7_5

study areas such as the Soviet atomic weapons programme.[4] British planners understood surveillance as ensuring coverage of strategically sensitive areas rather than panoptic viewing. US planners had a much grander prospect: that of encompassing everything from the sea floor to space in a reconnaissance network. Employing hydrophones, radar, spy satellites, seismographs, and various electronic surveillance techniques to intercept "enemy" communication networks (electronic signals intelligence, ELINT), the USA aimed to place the whole of earth under surveillance.[5] In the 1960s these competing perspectives on surveillance informed ongoing political debates and military plans pertaining to coordinated defence of key sea areas, encouraging joint monitoring between allies.

These contrasting surveillance strategies had a major impact on the trajectories of military oceanography in Britain. Greater knowledge of the physical properties of the oceans was of fundamental importance for defining the accuracy of the detection of enemy vessels in general, and submarines more specifically. Thus, the debate on alternative surveillance strategies informed key decisions regarding the NIO research programmes, and at the same time shaped the career of George Deacon.[6]

Competing Philosophies on Ocean Surveillance

The Cold War has often been described rather simplistically as a binary conflict between two homogeneous alliances, whose solidarity was maintained by mutual enmity towards the "other" bloc. In reality the Western alliance was far more fragmented.[7] At least up until the mid-1960s American and British military plans were based on diametrically opposed visions, and, as noted by commentators at the time, 'Unlike the United States, where military planning is predicted mainly on assessment of enemy *capabilities,* British planning gives somewhat greater weight to enemy *intentions*'.[8] Leading naval officers in the UK feared mainly airborne threats, especially nuclear missiles, and believed that sea surveillance should be kept to a minimum. In 1962 British Vice Admiral Varyl C. Begg aptly argued that: 'the facts of life are that to destroy this country overnight there would be no point in the enemy coming by sea!'[9]

That said, since control of communications across the Atlantic was going to play a key role in any future conflict, both the US and British navies sought to invest in electronic equipment securing sea communications and detection of enemy vessels. The advent of submarines carrying nuclear missiles led the US Navy to set up arrays of hydrophones far away from its country's coasts so as to offer advanced warning of imminent Soviet attacks on North America. Conversely, the British Admiralty had no intention of investing substantially in a detection system that offered no protection to Britain. Both countries had very different views on how electronic surveillance equipment would be used by their militaries.

These divergent positions had their origins in the formation of post-war military strategy. When the Second World War ended, British and US naval commanders had distinct ambitions with regards to ocean surveillance. When, in 1949, the North Atlantic Treaty Organisation (NATO) assigned to the Royal Navy responsibility for monitoring the Eastern Atlantic, the new task provided a lifeline to the costly force that in 1948 had been reoriented by defence chiefs towards anti-submarine warfare, although in 1957 Duncan Sandys still described the Royal Navy's future role as 'somewhat uncertain', and his White Paper argued for a reduction of future investment in Britain's naval force. However, the Royal Navy's responsibility to NATO in the Eastern Atlantic enabled the funding of novel ASW techniques and surveillance systems.[10]

By the late 1950s, the US Navy could lavish funds on very expensive pieces of ASW equipment and design multiple far-reaching projects to deploy surveillance virtually everywhere. Whilst the Royal Navy had seen deep funding cuts, the reality for the US Navy was the opposite. Predictably, during the early Cold War, the US Navy expanded rapidly overseas, establishing bases throughout the world often in direct competition with existing Royal Navy facilities. One notable case was the US naval station established in 1953 at Rota (Spain), just over one hundred kilometres north-west of Gibraltar. Moreover, the US Navy could also count on a constant flow of new oceanographic knowledge and the production of new technologies for sea surveillance through a well-oiled mechanism that secured regular disbursements from the US Office of Naval Research to a number of military and civilian research agencies and private companies.[11]

The differences between what the Americans and the British could afford became acutely apparent when plans for new sonar equipment were outlined. By the early 1950s both the US and UK navies had independent projects for the development of a passive sonar system.[12] Project Corsair, jointly undertaken by the British Admiralty Research Laboratory and its Underwater Detection Establishment, failed and was cancelled in 1957.[13] In contrast, its US equivalent, Project Jezebel, succeeded thanks to the involvement of an industrial partner, Bell Laboratories, which made the system more effective through the use of target identification equipment. Sonar came in two varieties—active and passive. In the Second World War active systems had been used, whereas in the post-war years more passive systems were deployed because submarines were now being used to hunt for other submarines and required a mechanism for remaining undetected whilst seeking out enemies. As has been shown earlier, passive systems were preferable to active systems because they did not emit a pulse and therefore remained undetectable, whereas active systems revealed the location of the hunter to the hunted. Due to its own expanding submarine fleet the US Navy opted for a passive system, making the surveillance silent and impossible to detect—a powerful weapon in itself. From the mid-1950s onwards the passive Sound Surveillance System (SOSUS) was deployed on the Eastern Atlantic seaboard and there were plans to deploy it in the Pacific as well.[14] Conversely, as a consequence of the Project Corsair failure, British attitudes to maritime surveillance changed considerably, leading to greater emphasis on developing a novel active sonar system instead.

This divergence emerged at a crucial point in time for Anglo-American relations. The years between the IGY and the Limited Test Ban Treaty (1963), when Britain, the USSR, and the USA agreed to ban atmospheric testing of nuclear weapons, were typified by mutual military–scientific interdependence in oceanographic research. This was precisely because of the growing need for reliable ocean surveillance as nuclear missile-carrying submarines became operational.[15] Variations in currents, temperature, salinity, and pressure (depth) caused phenomena known as thermoclines, which altered the characteristics of sound waves travelling through seawater. Thermoclines caused the path of sound waves to be bent rather than remaining linear. As a result submarines could hide in plain sight, invisible to sonar. Naval commanders viewed thermoclines alternatively as a defensive challenge or an offensive advantage, depending on whether they were trying to hide or seek. Nevertheless the pursuit of surveillance through passive and active sonar instigated collaboration between naval commanders and oceanographers, since marine scientists had already investigated waves, currents, and temperature variations in the Atlantic and other oceans in the previous decade. In fact they had been responsible for transforming oceanography into a Big Science by using expensive vessels and instruments for this purpose.[16]

The implementation of surveillance systems depended upon the study of oceanographic characteristics and the setting up of surveys enabling marine scientists to chart them. The Strait of Gibraltar was the principal area of attention of naval officers and oceanographers in Britain and the USA alike, because of the threat of new Soviet diesel-electric powered nuclear missile-equipped vessels that came into operation in 1956 and were known within NATO as Golf-class submarines.[17] This led them to join forces, producing oceanographic knowledge and competing in initiating new surveillance schemes and systems. Also in response to the new Soviet submarines the USA and UK collaborated to create their own submarine nuclear missile system known as Polaris, although this was only agreed to in 1962 and the British did not have any operational submarines until 1967.[18] Nevertheless, the development of ballistic missile-equipped submarines promoted naval operations from the rearguard to the vanguard of the Cold War, establishing a whole new genre of secondary deterrence.[19] This fundamentally changed the nuclear "game", increasing the relevance of the oceans, and in turn the science of oceanography, in shielding the West against this new security threat.[20]

Sea Lines of Communication: Surveillance of the Gibraltar Strait

In the mid-twentieth century Gibraltar was not a typical colonial possession, but a large, although not modern, overseas British naval base.[21] "The Rock", as it was known, had been in British hands since 1704, and the Royal Navy had permanently kept a contingent there. After the Second World War the naval base continued to play a role as a key communications and transit centre for the management of British interests and colonies abroad.[22] Vessels travelling to British colonies used the facilities at Gibraltar to refuel and make temporary repairs. The base provided significant geostrategic and geopolitical capital as well as justification for Britain's continued involvement in the Mediterranean, since its location allowed the monitoring of all shipping moving to and from the Atlantic.[23] British and American planners agreed that the sea areas adjacent to Gibraltar ought to be surveyed, but they had very different strategies for how these should be monitored.

A number of Cold War developments, including some close to Gibraltar, prompted the US Navy to challenge Britain's unilateral control of the Strait and intensify their presence there. The 1956 Suez Crisis heightened tensions in North Africa, while France and Spain were forced to withdraw from their African colonies shortly afterwards.[24] These withdrawals were the first major steps towards decolonisation and, with Western European forces retreating, the Soviet Union positioned itself as the bastion of international freedom and anti-colonialism in these regions.[25] The supply of arms and machinery to the North African states resulted in an increased number of Soviet freighters operating in the Mediterranean and close to Gibraltar.[26] The Soviet presence represented a threat to American and British hegemony in the Mediterranean and was a major catalyst for designing new surveillance plans.[27]

Surveillance at Gibraltar was primarily a concern of the US Navy ASW Director, Rear-Admiral Lawson P. Ramage, but the Royal Navy's Flag Officer (senior British officer) at Gibraltar was also preoccupied with the protection of communication lines near the Strait. Gibraltar would be the last line of detection (and defence if necessary) before Soviet submarines, leaving their base in the Black Sea, could reach the Atlantic.[28] In 1960 the Royal Navy's Director of Undersurface Warfare, George Symonds, reported on the Strait of Gibraltar's strategic role and recalled the threats to be derived from the recent construction of a Soviet base in the Mediterranean.[29] If Symonds agreed with Ramage on the nature of the Soviet threat, however, he disagreed on the exact shape and form of surveillance system to be deployed.

At the heart of the dispute were differing assessments of the cost-effectiveness, efficiency, and adaptability of competing systems. The American proposal, in line with the ongoing development of the SOSUS network, was to establish a chain of fixed passive sonar installations facing the Atlantic side of Gibraltar.[30] Symonds resisted this proposition and argued instead for an active sonar system. He believed active sonar was the best surveillance technology to be deployed at Gibraltar because the knowledge and expertise that Project Jezebel had given to the US Navy was not held within the British Admiralty or the Royal Navy. Furthermore, the Admiralty ceased research into fixed passive sonar arrays following the Project Corsair failure.[31] Finally Symonds believed it necessary to deploy a system capable, if necessary, of seeking and destroying nuclear submarines, especially since these would represent a direct threat to British territories.[32] Thus he suggested countering the US proposal with a British plan to produce helicopters fitted with deep-dangled sonar on a long cable to unite surveillance and offensive weaponry capable of destroying submarines.[33] The Admiralty had been developing the Wasp helicopter specifically for this role since 1958.

In order to verify the efficiency of the proposed ASW system, the Admiralty arranged a survey of the sea area adjacent to Gibraltar. In the summer of 1960 the Admiralty Underwater Weapons Establishment (AUWE), the Admiralty Research Laboratory, and the NIO jointly undertook the exercise with the assistance of NIO's research vessel *Discovery II*.[34] As shown in the previous chapter, Deacon had been concerned about the involvement of the NIO in other departments' major projects due to the failure of TRE equipment tested in collaboration with the Admiralty Research Laboratory in 1952. As a result of this he had looked for international collaborative work within the context of the IGY instead. The surveillance of Gibraltar presented him with the same dilemma, but obviously, given the significance of surveillance operations, he could do little to oppose the request for assistance from the other two defence research laboratories. Ultimately the objective of the survey was to gain a better understanding of acoustic propagation in the Strait's waters and testing signal strength at different depths to ascertain if the deep-dangled sonar would succeed.

US naval officers now sought to convince their allies in Britain about the urgent need to survey the Strait's waters, juxtaposing the Admiralty's project with one bringing together other US allies. Joint Canadian, US, and British oceanographic research had already taken place in Gibraltar in 1959.[35] Meanwhile NATO took responsibility for sponsoring collaborative oceanographic projects in light of ballistic missile-carrying submarines which entered service in both the Soviet (1956) and US navies (1959), and as discussions developed over the creation of a nuclear submarine-equipped pan-Western European multilateral force (MLF) to man a European secondary deterrence.[36] This compelled the alliance to secure a better understanding of key sea passages such as the Gibraltar and the Turkish Straits, and the Norwegian Sea. Through its Science Committee, NATO established a sub-group: the Sub-Committee on Oceanographic Research, responsible from 1960 for organising the surveys.[37]

Deacon became a key figure in this organisation, which was formed by some of the Western oceanographers who had been part of the IGY oceanographic committee (Danish marine biologist Anton Bruun, WHOI Director Columbus Iselin, Håkon Mosby of the Bergen Geophysical Institute, Henri Lacombe of the Paris Laboratory of Oceanography).[38] It is reasonable to assume that when plans for establishing the sub-committee were outlined Deacon hoped to play the same game he had played before,[39] namely to play national collaborative projects off against international ones, so as to decide contingently where more resources had to be placed. Additionally, the committee's agenda, to allow oceanographers in NATO countries to join forces in major surveys of key strategic areas, was consistent with his views on the purpose and direction of international collaboration in oceanography. As shown in this book and also discussed by Hamblin and others, the IGY had been extremely important to Deacon's career but had also made him aware of the implications of Soviet oceanographic work for defence and surveillance operations.[40] Through NATO sponsorship, Deacon hoped to re-create the same opportunities for collaboration that the IGY had bestowed, but more in line with existing political alliances.

By the end of 1960 both the US and British naval authorities were looking at NATO as the space where decisions about Gibraltar would be taken, but while the US Navy was lobbying for joint oceanographic exploration, the British Admiralty hoped to convince NATO allies to pay for the sonar to be made in Britain. However, the final report of the Admiralty survey showed that the acoustic properties of the sea bed required a more detailed study and that the lack of knowledge of Spanish territorial waters prevented the deployment of the new British device. Handling and controlling the equipment in the Strait's 'strong and variable tidal conditions' was going to be challenging.[41] Despite these limitations, Symonds now advocated the adoption of the deep-dangled sonar, arguing that it was going to be cheaper and more effective than the US-made sonar array. Moreover, Symonds argued that the Strait would be an important choke point in a future war; a purely defensive surveillance system would not work. If Gibraltar was now to become an "unofficial" NATO base, the Admiralty felt that all their allies should be compelled to contribute to its surveillance.[42] Symonds' summary thus launched the ambitious British project and urged NATO assistance: 'we cannot ignore the defence of the Straits of Gibraltar, but on the other hand we cannot "go it alone" and NATO countries must assist.'[43]

The 1960 UK Defence White Paper further restricted the budget for military research and gave the Admiralty limited funds for science projects. Soon after its release a discussion developed between the Admiralty's Director of Plans, Peter Ashmore, Director of General Weapons, Michael Le Fanu, and Symonds about the best use of these funds with regard to surveillance projects.[44] All three agreed that the construction of a surveillance system at Gibraltar had implications for wider NATO strategy and should not be solely paid for by the Royal Navy.[45] But since both the

Canadian and the US navies had shown no interest in the British system, the deep-dangled sonar scheme floundered in Admiralty bureaucracy. In 1961 even the Under-Secretary of State, Nigel Abercrombie, failed to find sufficient funds for full-scale trials.[46]

The Civil Lord of the Admiralty, Ian Orr-Ewing, who had been brought in by the Prime Minister, Harold Macmillan, to curb spending in the Royal Navy, now took responsibility for assessing the feasibility of the British project and requested that the Third Sea Lord, Peter Reid, prepare a report. Reid's document was strongly worded. SOSUS was described as being fantastically expensive and ineffective: 'The reason for our basic disagreement with the American concept of a static [anti-submarine] Defence of the Straits in peace time is that it is a purely global war requirement which, by government policy, is at the bottom of our list of priorities.'[47] Static defences were of 'no use either in peace or war unless there is a force of [anti-submarine] vessels…to classify and follow up the contacts which the static defences report.'[48] Reid convinced Orr-Ewing that the British active sonar represented the way forward, but when the Civil Lord of the Admiralty enquired again about NATO's European allies taking on some of the expense and research for the project, he found out that none was willing to do so.[49] The project thus stalled and it became apparent that placing the Strait of Gibraltar under surveillance had become even more urgent.

In 1963 French forces abandoned the Tunisian city of Bizerte, which hosted a French naval base and the facilities of the Allied Forces Mediterranean (AFMED).[50] As a result convoy routes in the southern Mediterranean that NATO forces patrolled and planned to use in time of conflict were now exposed to a greater risk. The NATO Commander-in-Chief Allied Forces Mediterranean, Vice Admiral A.B. Cole, claimed that the offensive capability to defend the eastern approaches to the Strait of Gibraltar was compromised. The Gibraltar naval base now had to project naval power in a 180-degree south-facing sweep. The Soviet threat had also increased due to submarine activities (potentially involving nuclear submarines) and 'hydrographic/ELINT' ships in the Mediterranean.[51] In that year one NATO report speculated that Soviet ships gathered both scientific data and intelligence, but in fact, little was known about the true purpose of these vessels. The memo underlined the effects of decolonisation, the rise of Soviet naval power, and the reduction of NATO military geographic deployment. Inevitably Gibraltar was to become the main NATO naval and military base and focus for offensive anti-submarine operations. The question remained, however, as to whether and how it should be patrolled.[52]

Cole, a New Zealander who had been in the Royal Navy since the end of the Second World War, vigorously argued for effective surveillance of the Gibraltar Strait in order to 'detect, identify and destroy transiting enemy submarines'.[53] He also emphasised the benefits to be derived from surveillance of merchant shipping transiting the Strait. This wider conception of the purpose of surveillance lent support to the British Admiralty's plans for a deep-dangled sonar, emphasising its role in peacetime surveillance. In a clear reflection of the Cuban Missile Crisis of the previous year, the report suggested that the surveillance system would provide 'useful intelligence information in peace or in times of tension'.[54] Cole also stated that although NATO had already spent considerable sums in improving the military strength of the naval base, there was no progress on securing 'comprehensive and effective surveillance system in the Gibraltar Straits'.[55] Attached to the memo was a report detailing operational requirements and suggesting that the British system was a cheaper alternative to be adopted in any case before a fully-fledged fixed installation could be completed.

Cole's appeal produced little in the way of results. The only item agreed upon at NATO level was to continue surveying the Strait, a project that found sympathy in some NATO countries such as France, Belgium, Norway, and Italy. The study of the formation of the layer of high salinity that travelled from deep Mediterranean waters in the Atlantic was of particular significance and had been studied by French oceanographer Henri Lacombe since the early 1950s when he had led a French research programme in the western Mediterranean.[56] In many ways, Lacombe can be considered the French equivalent of George Deacon. Described by the European envoy of the US Office of Naval Research as the 'rising star' of French oceanography, he had essentially taken the lead in NATO scientific exercises in the Mediterranean, whereas Deacon and Mosby had taken responsibility for those in the Northern Atlantic.[57] Lacombe, like Deacon, saw these exercises as continuing previous IGY work, and he organised the NATO western Mediterranean survey as a follow-up to France's participation in the IGY alongside Spanish oceanographers.[58] The presence of Spain in oceanographic expeditions represented a bone of contention for Deacon and British scientific groups, because of the political sensitivities involved. Conversely US authorities had no time for petty territorial disputes getting in the way of the need for comprehensive surveillance of

Soviet naval threats. In 1963 the US Navy intensified its efforts and sought now to 'investigate by analysis and/or model study the effect of currents on transducer orientation', while the US Office of Naval Research in London collected new data on subsurface currents in the Strait in order to improve the understanding of acoustic properties and to aid the design of a more effective passive system.[59]

Although there was cooperation on oceanographic work between the British and US navies, researchers had little to share in terms of what kind of surveillance Gibraltar really needed. British plans for 'on-the-spot' detection and response with ASW helicopters did not persuade the Americans, who—by contrast—looked at Gibraltar only as a knot in a vast network of SOSUS arrays extending from the seas surrounding Alaska to the Indian Ocean. These competing visions overlapped with competing strategies for diplomatic relations with countries neighbouring Gibraltar and in particular with Spain. Actually, it was to be Britain's changing relations with Franco's regime that sanctioned the final defeat of the British surveillance system and the adoption of an American one instead.

Decisions on the surveillance at Gibraltar did not depend solely on the availability of oceanographic data or the cost-effectiveness of new sonar technologies. Relations with Spain were important as well.[60] In the 1960s Franco's Spain was still isolated internationally and, although a member of the United Nations from 1955, it had no representation in NATO or any other pan-European organisation. Following the signing of the 1953 Pact of Madrid, the US administration had offered Spain a way out of isolation, mainly because of Spain's strategic positioning between the Atlantic, the Mediterranean, and North Africa.[61] One important consequence of the agreement was the offer of military aid in exchange for placing US naval and air bases on Spanish territory. As a consequence of these developments the US Navy could establish a station in Rota. When in the 1960s the NATO project for oceanographic surveys covering the Strait of Gibraltar and the adjacent Alboran Sea began, Lacombe's group and Belgian oceanographers invited their Spanish colleagues to join in. Consequently the Spanish vessel *Xauen* of Madrid's Oceanographic Institute could take part in the scientific exercise.

Conversely the NIO ended up playing a peripheral role despite the fact that Deacon had offered to lead the expedition. The reason for that has to be found in two equally important issues. Firstly, while the strategy of playing national projects against international ones had been effective thus far, this approach now stretched NIO resources too far. Since the NATO scientific exercise was to take place at the same time as the joint survey planned by the ARL and the Atomic Weapons Research Establishment (AWRE), Deacon had to make a decision about whether to opt for one or the other. Secondly, opting for participation in an international exercise including a Spanish contingent rather than a national one with implications for naval defence would have been unwise politically and would certainly ruin Deacon's relations at home. This may well have been the time that Deacon realised that keeping together strands of his policy network stretching nationally and internationally and comprising groups with competing agendas was extremely problematic.[62]

The British Admiralty had failed to convince the US Navy or a NATO ally to adopt the deep-dangled sonar, and Britain's role in NATO oceanographic efforts was being marginalised, with one Royal Navy official now turning to the old Spanish enemy for support. Presumably this officer believed that the US Navy would be more sympathetic towards the British sonar if the two European countries controlling the Strait agreed on the need for adopting it. So in 1963 the British commander in Gibraltar, Rear Admiral Errol Sinclair, arranged a private meeting with the Spanish Minister of Marine, Pedro Nieto Antúnez, to restore cordial relations, which had been strained for many years.

During the meeting the issue of securing a new surveillance system for Gibraltar was discussed together with other pressing issues including the presence of Soviet vessels in the Gibraltar Bay area and the possibility of joint scientific and naval exercises.[63] The decision to hold talks with a non-NATO ally on surveillance issues shows that the local British commander perceived coordination in the Strait area as urgent. To improve the chances of the new deep-dangled sonar, Sinclair was now even prepared to talk with representatives of a country that had been unfriendly up until that point. He even disclosed restricted information to Antúnez, eventually reporting to his superiors that he had just presented his personal views, 'coloured by his task at Gibraltar'.[64]

Sinclair and Antúnez mostly certainly did discuss sensitive issues. For instance, they considered how Soviet merchant vessels feigned machinery damage while secretly offering military equipment to Moroccan forces in their anti-colonial struggle against Spain.[65] Sinclair explained to Antúnez that Soviet trawlers covertly gathered intelligence and scientific data. He hinted at the Anglo-American surveillance plans, hoping that he could win the Spaniard's support for the new British surveillance scheme.[66] While analysing Spanish plans to adopt a fixed active surveillance installation in Ceuta (on the southern shore of the Strait), he attempted to "sell" the

British ASW helicopter surveillance technique, suggesting that the scheme was more in line with the financial capabilities of Spain. Helicopters were 'invaluable as weapons carriers' too.[67] Antúnez replied that as far as he was concerned there were no difficulties in going ahead with adopting a joint surveillance scheme. 'You may quote me on this as you wish,' he even added.[68] Notwithstanding Sinclair's breach of security, the Foreign Office's Permanent Under-Secretary showed enthusiasm for his initiative, which finally seemed to cast some positive light on the future of the deep-dangled sonar. 'Apart from the defence aspect, co-operation between the respective naval authorities may help to remove the political difficulties between Britain and Spain in relation to Gibraltar,' he argued.[69]

Nevertheless in 1965 the Royal Navy still unsuccessfully attempted to establish an Anglo-Spanish agreement on cooperation in oceanography.[70] In the same year Spain requested that Gibraltar be returned to the Iberian country at the UN's Anti-Colonisation Committee. A Red Book published by the Spanish Foreign Ministry now openly divulged that Britain had no interest in the Gibraltarians and planned to keep "the Rock" mainly because it needed a military base from which to track Soviet rockets and submarines. The recent secret conversation between Sinclair and Antúnez clearly proved it.[71]

The new Labour government in London elected in October 1964, under the leadership of Harold Wilson, now shifted firmly away from co-operating with Franco. In 1966 Spain's government retaliated by placing restrictions on British aircraft crossing Spanish airspace when flying towards Gibraltar. These restrictions later extended to motor vehicles crossing the border. The British government needed to counter the Spanish allegations that they only kept Gibraltar for its surveillance implications, organising a referendum. The day before the referendum NATO conducted naval exercises launched from Gibraltar which, unsurprisingly, excluded Spanish naval forces.[72] In 1969, as tensions escalated, Franco's government chose to close the border, and it remained closed until well after his death in 1975. By then, however, the situation with regard to the surveillance of the Strait of Gibraltar had rapidly evolved in favour of the US proposal. Since Britain had failed to find countries—within and outside NATO—sympathetic to its surveillance project, it was mothballed. Conversely by 1974 US plans for establishing a world-wide sea surveillance network, including a fixed installation on the Atlantic side of Gibraltar, had reached completion. In the end the British Admiralty had to agree with the US plans for fixed installations since an alternative scheme no longer existed.[73] Even more worryingly for the Admiralty, Gibraltar was not going to be the only case when British surveillance plans were cast aside in favour of more ambitious and expensive US surveillance projects.

Global Seas, Global Surveillance: The GIUK Gap

The oceanographic surveys carried out in the context of NATO-sponsored research aided in addressing the problem of how to improve detection in the presence (or absence) of currents and other oceanic conditions in Gibraltar. In the North-east Atlantic the confluence of cold Arctic water and warm Atlantic water in the Southern Norwegian Sea had been a subject of enquiry for oceanographers since the early twentieth century. A lot more needed to be understood about this convergence, especially if Western naval forces could efficiently monitor this strategically vital chokepoint that granted the Soviets access to the Northern Atlantic.[74] Unsurprisingly, the USA–UK dispute on competing surveillance schemes that typified Gibraltar continued in the cooler Atlantic waters off the Norwegian coast. This time, however, British officers were especially worried about the somewhat forceful way in which their US colleagues sponsored NATO oceanographic surveys in line with their ambitions for ensuring global coverage in surveillance operations.

In 1960 the NATO Sub-Committee on Oceanographic Research planned further surveys focusing this time on the Faroe–Shetland Channel. The Committee's chairman, Norwegian oceanographer Håkon Mosby, took responsibility for leading the NATO expeditions in these strategically vital sea areas in order to produce novel data on salinity, temperature, and currents.[75] The NATO study was openly geared towards assisting in development of the SOSUS network as plans to extend it to the Northern European coastal line existed. There was no opposition in Britain to these plans, and the NIO was actually amongst the leading oceanographic institutions in the endeavour. During the expeditions, the NIO's vessel *Discovery* carried the Swallow Float, the new device developed by oceanographer John Swallow in an attempt to take current measurements at pre-determined depths.[76] The study produced novel data on thermoclines and allowed for the production of synoptic charts illustrating how water temperature changed in the sea channel (and thus affected sonar detection).[77]

By pushing NATO towards studies of strategic waters, the US Navy wished to generate consensus on its plans for sea surveillance and address—with the assistance of new oceanographic data—the limitations of the SOSUS technology. However, NATO's oceanographic research actually had a very disruptive effect on other oceanic surveys that had no surveillance ambitions, something that worried Deacon and other British government officials. Not indifferent to what had happened in Gibraltar, Deacon now feared that the NIO staff were overstretched in dealing with several national and international projects at once. Furthermore, he worried about rival international research schemes competing with the NATO effort. His chief concern was that this rivalry could compromise the precarious balance that had enabled him to become the figurehead for British oceanography at home and abroad. This is exactly what happened, due especially to the dissatisfaction of marine scientists in Britain. In 1960 the International Council for the Exploration of the Seas (ICES) executed the Overflow expedition, aiming to chart oceanic characteristics of the Faroe–Shetland channel, the very same sea area now chosen by NATO.[78] ICES was primarily interested in fishery and collected, in contrast with NATO, both physical and biological data (including plankton sampling).[79] Crucially the USA was not an ICES member, while the Soviet Union was represented in the Council.

The planning of the NATO survey brought the collaboration between some European oceanographers that had been part of ICES to an end—at least for the next ten years—and strengthened the collaboration with US colleagues. Figures such as Deacon and Mosby, who had manifested some dissatisfaction with the ICES expedition, now took the opportunity to promote alternative NATO plans.[80] So, after taking part in the Overflow expedition, NIO's *Discovery* ceased to further participate in the ICES programme. Likewise, Mosby and the Norwegian oceanographers in the *Helland-Hansen* (which had also taken part in Overflow) now agreed to more surveys under the NATO aegis. This was in line with the stance of the Norwegian government that surveillance in the North Sea needed to be improved. Other experts involved in the ICES exercise and interested primarily in fishery studies did not welcome NATO's intervention in oceanographic affairs. The "invasion" of NATO oceanographers into a delicate environment for fisheries research caused outrage amongst scientists which rumbled on in Britain for the next five years.[81] In particular, J.B. Tait of the Torry Marine Laboratory in Aberdeen continued to be part of the NATO survey until 1962, but after that the he became a staunch critic of NIO's director.

Deacon had been the man of compromise finding a space for biologists in the NIO, but the debates around NATO-sponsored research in the North Atlantic undermined his ability to appease marine scientists. This being said, NIO's contribution to the Faroe–Shetland channel project also waned somewhat despite the initial enthusiasm shown by Deacon for taking it to completion. The chief reason for the diminished interest in carrying it forward was once again the criticism put forward at UK Cabinet level of the NATO work's disruption of traditional fishery research. When the Working Group on Oceanography of the UK Cabinet Committee on International Co-operation met on 3 September 1962, a MAFF representative argued that 'NATO studies…were *likely to give offence* to the USSR and jeopardise Russian collaboration in the field of fishery conservation'. Deacon replied at the meeting that: 'he saw the NATO project as supplementing fisheries research in the area, not as duplication'.[82] Although no official decision was taken at Cabinet level to address the resentment of fishery scientists, the contribution to NATO surveys of British oceanographic groups was affected by the altercation. Deacon's position continued to be strong overall but had now been undermined by the enmity of fisheries scientists in his network.

Meanwhile, US defence planners capitalised on the NATO studies and swiftly moved from the stage of oceanographic research to that of setting up new sonar installations. The Cuban Missile Crisis of October 1962 revealed that monitoring Soviet submarines from the Eastern Atlantic seaboard was insufficient. If tracking was to be effective throughout the North Atlantic, then SOSUS required extension.[83] The tracking of Soviet merchant vessels carrying nuclear missiles and launching equipment to Cuba had revealed that another threat existed aside from submarines. The SOSUS system deployed off the Florida coast was effective but could not ensure coverage of other Atlantic sea areas. In particular the sonar array ceased to offer clear detection when climatic conditions changed around the mid-Atlantic ridge.[84] These changeable conditions, therefore, dictated the maximum range of detection daily. In the months following the Cuban Missile Crisis it became apparent to US Navy officials that a second line of subsurface sonar stations in the North-East Atlantic would have to be given greater priority and the installation programme was accelerated.

When the NATO surveys ended, a wealth of new oceanographic data was made available to Western navies' officers. Those of the US Navy now agreed to discuss with the British Admiralty the possibility of improving the efficiency of SOSUS in the Atlantic. The conversation was obviously facilitated by the conclusion of the dispute in Gibraltar and the American officers knew that since the Admiralty no longer intended to sponsor its sonar system in the southern European Strait, it could be persuaded to adopt the American one further north. In June 1967 a small team from the Admiralty Underwater Weapons Establishment (AUWE) was invited to the USA.[85] During the trip the Admiralty envoys observed a SOSUS facility and the US Navy officials even proposed that innocuous cover names could be adopted. This would make it easier to convince outsiders/enemies that the facility's 'activities are concerned entirely with Oceanography'.[86] The AUWE team now realised that the offer was being made because the central Atlantic ridge seriously affected detection performance. Setting up a SOSUS station in Britain could save the US Navy up to US$19 million by enabling them to relay and route the signals differently.[87] Weighing the pros and cons of UK participation, the AUWE officers reported that the scheme was beneficial to Britain's defence and foreign policy objectives.[88]

After so many disputes with the Americans, the new project for a SOSUS station in Britain restored a "special" relationship on surveillance affairs. Yet, in casting a positive light on the collaborative deal, the Admiralty team's report now suggested that the British active sonar scheme proposed in previous years was not *alternative* to the SOSUS but actually *complementary*. 'Collaboration in SOSUS would be concrete evidence of sincerity in collective ASW defence', it argued, and 'the large UK investment in ASW frigates, helicopters, submarines and aircraft could be put to effective use.'[89] In fact, the Admiralty document indicated, it was *exactly* because of SOSUS that the British investment in sea surveillance could be made cost-effective: 'it would significantly improve the UK military control…without incurring large overseas expenditure from forces abroad.'[90] The only problem was 'the effort involved in "Anglicisation" of US equipment'.[91]

Despite this, the project was given the go-ahead.[92] Project Backscratch (the cover name for the construction of a UK SOSUS station) was meant to include a British contribution of £5.6 million (mainly for costs of surveys and providing shore facilities). The US Navy provided all arrays, cables, and hydrophones.[93] The Royal Navy had to be careful since it was about to close overseas bases to fund its existing projects and the endorsement of a costly Anglo-American scheme seriously impacted on the management of ongoing programmes abroad.[94]

In 1970 the SOSUS project was revisited as the new Conservative Prime Minister Edward Heath entered office. The original commitment of £5.6 million had by then been significantly revised. Although there was still to be a full intelligence-sharing agreement, the Ministry of Defence was now required to provide the building to house the shore station to be built at the Royal Air Force facility of Brawdy.[95] The site was chosen because it allowed US planes to land and resupply the base. It also suited British ASW plans. The new Defence Secretary, Peter Carrington, presented these plans in the Cabinet meeting of 28 December 1970. Britain would pay only £1.5 million, a small fraction of the US funding of SOSUS (£80 million) and yet still a significant contribution to a project that did not align with Britain's approach to surveillance.[96]

The facility at RAF Brawdy came into service in 1974 and remained operational until 1995, when it was finally decommissioned. Anti-nuclear campaigners did not welcome the facility; they believed it hosted a nuclear missile command centre, and it was even protested during the 1980s by the women of Greenham Common. In reality it was staffed by US Navy officials operating the SOSUS detection equipment. Much like Gibraltar, its significance was not in fixed installations at sea, monitored from Brawdy, but rather in the integration of signals from any remote place in the world that could be picked up and transmitted to other knots in the SOSUS network. Brawdy, as many other similar bases around the world, demonstrated that American plans for global sea surveillance had proven successful precisely within a country where they had initially been firmly opposed.

Ocean Surveillance and Oceanography

Speaking at the 1982 conference on the implementation of the United Nations Convention on the Law of the Sea (UNCLOS), the then Director of the Institute of Oceanographical Studies (as the NIO had by then become), Anthony Laughton, reflected on the chief aspects of oceanographic research in the 1960s:

> Most oceanographic research has a complex mixture of motives and funding, sometimes related to the long-term needs of defence, sometimes to the needs of scientific curiosity, sometimes to the assessment of potential

resource exploitation, sometimes commissioned by government or industry to achieve defined objectives related to the utilization of the oceans.[97]

In separating the various motives for funding oceanographic research Laughton did not mention the use of oceanographic science as a means for surveillance and intelligence gathering, or the significant investment in technologies securing surveillance of the sea. Military support for oceanography, which had led to the rapid expansion of the discipline during the 1950s and in particular the 1960s, was premised on the fear that the sea had hidden threats deriving especially from the movement of enemy submarines. This fear continued to blur the boundaries between military operations and oceanographic work even when the systems of surveillance discussed in this chapter were finally operational. In 1968 the USS *Pueblo* was captured by communist North Korea whilst carrying an assortment of ELINT and SIGINT equipment, when the vessel was supposedly carrying out oceanographic survey work.[98] For Western navies the rapid expansion of a Soviet oceanographic research fleet chimed with continued fears about the activities of Soviet trawlers operating off the coasts of Scotland, Iceland, and the Mediterranean carrying out covert surveillance operations.[99]

Oceanographers and naval officers jointly pursued sea surveillance within the context of the special relationship between US and UK navies and the NATO alliance. The two nations had different plans for enforcing and deploying surveillance of the sea. The United States saw this problem as a global one. Its fiscal supremacy enabled costly projects to be implemented throughout the world's oceans. Of the three military surveillance technologies adopted in the last century—sonar, radar, and satellite—the first remains the most secretive. The case of the United Kingdom is quite different. Here the development of ocean surveillance has to be seen in the light of an increasingly restricted world presence, limited budgets, and a geographical location. For the Royal Navy the threat was from Soviet submarines striking a conventional attack against its home bases. So the Royal Navy saw that standing alongside its American counterparts was important, but not a carte blanche to squander precious resources of its own.

By looking at both the case of Gibraltar and the GIUK, it can be seen how the British position changed considerably over time, partly because of lobbying by the US Navy for the adoption of SOSUS, and partly due to the financial difficulties of developing a viable British alternative to the US sea surveillance system.

The early proposal for the active sonar system in Gibraltar found little support amongst NATO allies and was quietly dropped. It was reconsidered in light of an improbable alliance with Spain and was eventually abandoned. After that, the US Navy gave the Admiralty the opportunity to prevent criticism at home by finding a different use for the sonar made in Britain that never materialised in Gibraltar. By proposing to set up a SOSUS station in Wales as part of a major scheme to better reconnoitre the North Sea, the US Navy effectively provided the Admiralty with an opportunity for utilising the otherwise unusable British surveillance equipment.

On the other hand, the differences that typified British and US views on what surveillance meant and what one could aim to achieve with it continued. The problem was not, of course, just about sea and its invisible threats, but more generally about the changing role that Britain and the USA were about to play on the geopolitical chessboard of the Cold War. Approaches to surveillance in Britain were consistent with the geopolitical ambitions of a declining colonial power that had lost most of its possessions abroad and wished therefore to exclusively monitor what threatened its national security in selected parts of the world; incidentally those strategic bases retained after the Empire had collapsed. Conversely, the USA sought to reaffirm its role as Cold War hegemon by investing significantly in technologies that would ensure a global vision on existing threats (at sea and elsewhere).[100] The investment was consistent with a geopolitical stance that recognised that territorial occupation was a far less effective means of ensuring worldwide domination than securing the surveillance of distant places. The new American empire, in contrast with the old British one, was managed through the placing of invisible "electronic ears" underground, on land, at sea and in space all around the globe.

Despite the disagreement between the US and UK naval planners, in both countries (and NATO more generally) naval officers in charge of surveillance programmes understood the importance knowledge of the characteristics of the ocean had for the pursuit of sea surveillance. Oceanographers were thus mobilised to survey those sea areas that needed routine reconnaissance as lack of knowledge of sea physical characteristics could jeopardise these patrolling operations. Throughout the period British and American oceanographers worked jointly on these areas, and both civilians and military scientists collaborated. From 1963 the US

Office of Naval Research gathered all known data from European sources on the chokepoints at Gibraltar and the Faroe–Shetland channel, making use of the data-processing facilities available in the USA, and started to complete the panoptical study of oceanic waters surrounding the continents in the northern hemisphere.

The offer of collaboration in oceanographic enterprises was also an effort to dissuade the British Admiralty from adopting its own surveillance scheme, which greatly reduced the need for oceanographic investigations of military significance in Britain. This did not immediately result in the decline of George Deacon as network builder. In fact, during these years he was at the height of his powers, serving as a civilian on international committees of both military and civilian composition. Utilising the international dimension, he succeeded in strengthening his domestic position as head of the NIO. But it is equally true that he created enmity between himself and the fisheries scientists because of his collaboration in NATO.[101] This collaboration jeopardised links between British and Soviet researchers at a time when the global conservation of fish populations was an emerging diplomatic issue.[102] Moreover, Deacon found it increasingly difficult to have the NIO contributing to both national and international projects. When he made decisions about what to prioritise, he weakened the links either with his connections at home or with those abroad.

Deacon's domestic network, whilst initially strengthened though international collaboration, was eventually eroded due to challenges to his leadership from within Britain, as the number of ocean scientists from within and outside the NIO community who took issue with the extent of his power grew. Thus, when the NIO activities were reviewed in the mid-1960s, some of Deacon's opponents openly voiced their concerns, making it more difficult for him to continue building his network. In fact, this criticism was decisive in supporting the position of those, within government and the Admiralty, who saw the hybrid military/civilian model pioneered by the NIO as increasingly problematic. This criticism reflected changing perceptions regarding the role of oceanographic research in naval operations in general and surveillance more specifically. With the Admiralty no longer in a position to develop a British-made surveillance system, the interest in an oceanographic establishment geared towards military oceanography waned somewhat. As the next chapter shows, Deacon eventually had to succumb, as the Admiralty withdrew the support that made it possible for him to either ignore or appease his rivals.

Notes

1. Report for the Civil Lord, 17 August 1961, ADM 1/29275, TNA (London).
2. Kirstie Macrakis, "Technophilic Hubris and Espionage Styles during the Cold War," *Isis* 101:2 (2010): 378–385.
3. Signals Intelligence (SIGINT) is the interception of communications (signals), whether between people, or in the form of electronic signals not directly used in communication (see note 5.).
4. Richard Aldrich, "British Intelligence and the Anglo-American 'Special Relationship' during the Cold War," *Review of International Studies* 24:3 (1998): 331.
5. ELINT or electronic signals intelligence refers to the monitoring of signals not used for communications.
6. As this book shows, Deacon repeatedly returned to areas of oceanographic study that concerned the detection of vessels, from his war work (Chap. 2) to his formation of the research priorities of the institute (Chap. 3) to the use of acoustic instruments during the IGY (Chap. 4).
7. See Jeffrey A. Engel, *Cold War at 30,000 Feet: The Anglo-American Fight for Aviation Supremacy* (Cambridge, MA: Harvard University Press, 2007): 17–52; Alan P. Dobson, *Anglo-American Relations in the Twentieth Century: Of friendship, conflict and the rise and decline of superpowers* (London: Routledge, 1995).
8. W.P. Snyder, *The Politics of British Defence Policy* (Athens, OH: Ohio University Press, 1964): 29 [my italics].
9. Vice Chief of the Naval Staff to Flag Officer (Senior Royal Navy Commander), Scotland, 15 August 1952, ADM 1/28093, TNA (London).
10. Eric Grove, *Vanguard to Trident: British Naval Policy since World War II* (Annapolis: Naval Institute Press, 1987): 211. See also M. Dockrill, *British Defence since 1945* (Oxford: Wiley-Blackwell, 1988): 68.
11. On ONR funding to US oceanographic institutions, see Gary E. Weir, *An Ocean in Common. American Naval Officers, Scientists, and the Ocean Environment* (College Station: Texas A&M University Press, 2001) and Jacob D. Hamblin, *Oceanographers and the Cold War: Disciples of Marine Science* (Seattle: University of Washington Press, 2005).
12. Stephen Twigge, Edward Hampshire, Graham Macklin, *British Intelligence Secrets, Spies and Sources,* (London: TNA, 2009): 157.
13. Following the Portland Spy Scandal this establishment was consolidated into the larger, Admiralty Underwater Weapons Establishment (AUWE), which combined the Underwater Detection Establishment, Torpedo Experimental Establishment, Underwater Launching Establishment, in

1961. See Tom Wright, "Aircraft Carriers and Submarines: Naval R&D in Britain in the Mid-Cold War", in *Cold War Hot Science: Applied Research in Britain's Defence Laboratories 1945–1990*, ed. Robert Bud and Phillip Gummett (London: Science Museum, 1999): 153–4.

14. Jeffrey T. Richelson and Desmond Ball, *The Ties That Bind: Intelligence Cooperations between the UKUSA Countries- the United Kingdom, the United States of America, Canada, Australia and New Zealand* (London: Unwin Hyman, 1990): 200.
15. Hamblin, *Oceanographers and the Cold War*, xvii-xix. See also Simone Turchetti, "Sword, Shield and Buoys: A History of the NATO Sub-Committee on Oceanographic Research 1959–1973," *Centaurus* 54:3 (2012): 205–231. On the test ban treaty see Vojtech Mastney, "The 1963 Nuclear Test Ban Treaty: A Missed Opportunity for Détente?," *Journal of Cold War Studies* 10:1 (2008): 3–25; Glenn T. Seaborg, *Kennedy, Khrushchev and the Test Ban* (Berkeley: University of California Press, 1981).
16. Hamblin, *Oceanographers and the Cold War*, 264–5. On big science, see the introduction and Peter Galison, "The Many Faces of Big Science," in Peter Galison, Bruce Hevly, *Big Science: The Growth of Large Scale Research* (Stanford: Stanford University Press, 1992).
17. Norman Polmar and K.J. Moore, *Cold War Submarines: The design and construction of US and Soviet Submarines* (Washington: Potomac Books, 2004): 107–111.
18. Graham Spinardi, *From Polaris to Trident: The Development of US Fleet Ballistic Missile Technology* (Cambridge: Cambridge University Press, 1994); Maggie Mort, *Building the Trident Network: A Study of the Enrolment of People, Knowledge and Machines* (Cambridge, MA: MIT Press, 2002); Richard Crossman, *The Diaries of a Cabinet Minister: Volume Two* (London: Hamish Hamilton & Jonathan Cape, 1976): 619–620; Wright, "Aircraft Carriers and Submarines,"147–183.
19. Sherry Sontag, Christopher Drew, Annette Lawrence Drew, *Blind Man's Bluff: the Untold Story of Cold War Submarine Espionage* (London: Hutchinson, 1999): 46–53; Desmond Ball, "Nuclear War at Sea," *International Security* 10:3 (1985) 3–31; Barry Nalebuff, "Brinkmanship and Nuclear Deterrence: The Neutrality of Escalation," *Conflict Management and Peace Science* 9:2 (1986): 19–23.
20. Michael A. Dennis, "Earthly Matters: On the Cold War and the Earth Sciences," *Social Studies of Science* 33:5 (2003): 811–812; Naomi Oreskes, "A Context of Motivation: US Navy Oceanographic Research and the Discovery of Sea-Floor Hydrothermal Vents," *Social Studies of Science* 33:5 (2003): 697–742.

21. Peter Gold, *A Stone in Spain's Shoe: The Search for a Solution to the Problem of Gibraltar* (Liverpool: Liverpool University Press, 1994): 3–4; Kent E. Calder, *Embattled Garrisons: Comparative Base Politics and American Globalism* (Princeton: Princeton University Press, 2010): 11.
22. Jeremey Black, *The British Seaborne Empire* (London: Yale University Press, 2004): 242 and 356; Tillman Nechtman, "'...for it was founded upon a Rock': Gibraltar and the Purposes of Empire in the Mid-Nineteenth Century," *The Journal of Imperial and Commonwealth History* 39:5 (2011): 749–770.
23. Donald C. Watt, *Succeeding John Bull: America in Britain's Place 1900–1975* (Cambridge: Cambridge University Press, 1984): 12; Melvyn P. Leffler, *A Preponderance of Power: National Security, the Truman Administrations, and the Cold War* (Stanford: Stanford University Press, 1992): 123, 171, 273.
24. Egya N. Sangmuah, "Eisenhower and Containment in North Africa, 1956–60," *Middle East Journal* 44:1 (1990): 76–91; Matthew Connelly, "Rethinking the Cold War and Decolonization: the Grand Strategy of the Algerian War for Independence," *International Journal of the Middle East* 33:2 (2001): 221–245. See also Peter Hahn, *The United States, Great Britain, Egypt, 1945–1956: Strategy and Diplomacy in the Early Cold War* (Chapel Hill: University of North Carolina Press, 1991); Tony Chafer, *The End of Empire in French West Africa: France's Successful Decolonization?* (Oxford: Berg, 2002).
25. John Lewis Gaddis, *Strategies of Containment: A Critical Appraisal of American National Security Policy during the Cold War* (Oxford: Oxford University Press, 2005): 179–180.
26. Record of a meeting between Flag Officer, Gibraltar, and Spanish Minister of Marine, 2 April 1963, FO 371/169501, TNA (London).
27. Ennio Di Nolfo, "The Cold War and the Transformation of the Mediterranean," in *The Cambridge History of the Cold War*, ed. Melvyn P. Leffler and Odd Arne Westad (Cambridge: Cambridge University Press, 2010): Vol. 2, 238–57.
28. Harriet Critchley, "Polar Deployment of Soviet Submarines," *International Journal* 39:4 (1984): 836–7.
29. Note by George Symonds, Director of Undersurface Warfare, 21 December 1960, ADM 1/29275, TNA (London).
30. There are no official government documents that have been declassified which confirm its existence. However, several works claim that a SOSUS line west of Gibraltar exists: John Craven, *The Silent War: The Cold War Battle Beneath the Sea* (New York: Touchstone, 2002) 93; Peter Huchthausen and Alexandre Sheldon-Duplaix, *Hide and Seek: The Untold Story of Cold War Naval Espionage* (London: John Wiley, 2009): 113;

W. Craig Reed, *Red November: Inside the Secret US-Soviet Submarine War* (London: Harper Collins, 2010): 253–4; Sherry Sontag, Chistopher Drew, *Blind Man's Bluff: The Untold Story of American Submarine Espionage* (London: Harper Collins, 1998): 40–1, 68, 122; Robert E. Harkavy, *Bases Abroad: The Global Foreign Military Presence* (Stockholm: SIPRI, 1989) 193–4; William E. Burrows, *Deep Black: Space Espionage and National Security* (New York: Random House, 1988): 177–181; Jeffrey T. Richelson, Desmond Ball, *The Ties That Bind: Intelligence Cooperations between the UKUSA Countries- the United Kingdom, the United States of America, Canada, Australia and New Zealand* (London, Unwin Hyman, 1990): 198–202.

31. The Admiralty however continued to promote the study of passive sonar systems. It simply halted work on arrays of hydrophones. The demise of Project Corsair was divulged to non-scientific staff in the Admiralty only in 1968.
32. "Any form of detection device needs a weapon to back it up." Note by George Symonds, Directory of Undersurface Warfare, 21 December 1960, ADM 1/29275, TNA (London).
33. The system was known as ASDIC; Straits of Gibraltar Anti-Submarine Group, Minutes of Meeting held in Washington, 5 February 1960, ADM 1/29275, TNA (London).
34. Flag Officer, Gibraltar, to Commander-in-Chief, Mediterranean, 11 July 1960, ADM 1/29275, TNA (London).
35. In 1959 the CANUKUS group (made up of the three main NATO powers: Canada, the UK, and the US) commissioned a study of surveillance measures in Gibraltar; Minutes of Informal Meeting in Washington, 5 February, 1960, ADM 1/29275, TNA (London).
36. James B. Soloman, *The Multilateral Force: America's Nuclear Solution for NATO (1960–1965)* (Annapolis: US Naval Academy, 1999).
37. Turchetti, "Sword, Shield and Buoys," 214.
38. For more on the formation of the NATO sub-committee, see Turchetti, "Sword, Shield and Buoys," 211–212.
39. Turchetti, "Sword, Shield and Buoys".
40. Hamblin, *Oceanographers and the Cold War*, 66, 80, 88.
41. Captain Superintendent, AUWE to George Symonds, 23 November 1960, ADM 1/29275, TNA (London).
42. See Memo by A. G. Draper, Military Branch II, and Admiralty for T.A.K. Elliott, Foreign Office, 16 March 1964, ADM 1/29275, TNA (London); and Statement by Sir B. Burrows at the North Atlantic Council on 29 March, 1968, [available at www.gib-action.com/docs/nato1968.htm].

43. Note by George Symonds, Director of Undersurface Warfare, 21 December 1960, ADM 1/29275, TNA (London).
44. Correspondence by Director of Plans, 3 January 1961; Director of General Weapons, 3 February 1961; Director of Undersurface Warfare, 20 March 1961, ADM 1/29275, TNA (London).
45. Note, Director of Undersurface Warfare, 20 March 1961, ADM 1/29275, TNA (London).
46. Note by Nigel Abercrombie, USS, 26 May 1961, ADM 1/29275, TNA (London).
47. Admiral J. Peter L. Reid, Report for the Civil Lord of the Admiralty Ian Orr-Ewing, 17 August 1961, ADM 1/29275, TNA (London).
48. Ibid.
49. Note from Department of National Defence Royal Canadian Navy communicated through the Senior Naval Liaison Officer (British Defence Liaison Staff), 23 March 1962, ADM 1/29275, TNA (London).
50. Memo by Vice Admiral A.B. Cole, Chief of Allied Staff – NATO Headquarters, Allied Forces Mediterranean, Malta, 23 April 1963, ADM 1/29275, TNA (London).
51. On ELINT trawlers see Harkavy, *Bases Abroad*, 208–10.
52. Memo from NATO Headquarters, Allied Forces Mediterranean, Malta, (sgn. Vice Admiral, A.B. Cole), 23 April 1963, ADM 1/29275, TNA (London).
53. In the "Royal New Zealand Navy", at: http://nzetc.victoria.ac.nz/tm/scholarly/tei-WH2Navy-c29.html. Cole had served during the Second World War as commander of an anti-submarine destroyer. During the 1950s he rose within the Admiralty, serving as Assistant Chief of Naval Staff 1959–62 before taking up a NATO role.
54. Memo from NATO Headquarters, Allied Forces Mediterranean, Malta, (sgn. Vice Admiral, A.B. Cole), 23 April 1963, ADM 1/29275, TNA (London).
55. Ibid.
56. J. Bourgoin, "Henri Lacombe (33), 1913–2000. In Memoriam", *La Jaune et la Rouge* (February 2001): 562.
57. Simone Turchetti, "Sword, Shield and Buoys: A History of the NATO Sub-Committee on Oceanographic Research 1959–1973," *Centaurus* 54:3 (2012): 212.
58. Henri Lacombe, "Nato and Oceanography. Statement by Professor Lacombe", NATO, Annex H to AC/141-R/6: 3.
59. Memo from NATO Headquarters, Allied Forces Mediterranean, Malta, April 23, 1963 (sgn. Vice Admiral, A.B. Cole), ADM 1/29275, TNA (London).

60. This remains the case; see Sam Robinson and Lino Camprubí, "Ocean Science, Gibraltar and geopolitics – then and now," Science Blogs – Political Science (blog), *the Guardian,* http://www.theguardian.com/science/political-science/2013/oct/04/ocean-science-gibraltar-geopolitics/.
61. For a general discussion of Franco's role in sharing science and power, see Nestor Herran and Xavier Roqué, "An Autarkic Science: Physics, Culture and Power in Franco's Spain," *Historical Studies in the Natural Sciences* 43:2 (2013): 202–235.
62. Letter from Amirante Jefe del E. M. de la Armada to Sub-secetario del Ministerio Asuntos Exterores, on "Investigaciones Oceanográficas en el Estrecho de Gibrlatar en colaboración con Francia en su aspecto military", Madrid, 12 May 1958, MAE, Leg. R. 5334, Exp.20.
63. Record of a meeting between Flag Officer, Gibraltar, and Minister of Marine, Spain, 2 April 1963, FO 371/169501, TNA (London); Foreign Office's earlier stances are discussed in the file, "US & Spanish attitudes to Gibraltar 1952," FO 371/102020, TNA (London).
64. Record of a meeting between Flag Officer, Gibraltar, and Minister of Marine, Spain, 2 April 1963, FO 371/169501.
65. Lino Camprubí and Sam Robinson,"A Gateway to Ocean Circulation: Surveillance and Sovereignty at Gibraltar," *Historical Studies in the Natural Sciences* 46:4 (2016) 429–459.
66. Ibid.
67. Ibid.
68. Ibid.
69. Note by of N.J.A. Cheetham, FO Permanent Undersecretary, 2 April 1963, FO 371/16950, TNA (London).
70. See CO 926/2061, TNA (London).
71. Fernando María Castiella [Spanish Minister of Foreign Affairs], "Borrador de la introducción del Libro Rojo", April 1965, ACA, Leg. 69 Rel. 1 n15/19: 57.
72. Fernando María Castiella to Spanish Minister of Navy, San Sebastián, 26 August 1967, ACA, Leg 70. Rel 1. 15/19. 2, Carpeta 2. 213/220.
73. Harkavy, *Strategic Basing*, 127–129.
74. On this see Gunnar Ellingsen, "Varme havstrømmer og kald krig: 'Bergensstrømmåleren' og vitenskapen om havstrømmer fra 1870-årene til 1960-åreme [Warm ocean currents and the cold war: the Bergen current meter and the science of ocean currents from the 1870s to the 1960s]" (PhD dissertation, Universitetet I Bergen, 2012).
75. Turchetti, "Sword, Shield and Buoys," 215.
76. On Swallow see W. John Gould, "From Swallow Floats to Argo – The Development of Neutrally Buoyant Floats," *Deep-Sea Research* 52:3

(2005): 529–543 [available at: www.argo.ucsd.edu/Gould_Float_history.pdf, accessed 10/09/2013]. See also John Swallow, interview by Margaret Deacon, 1994 [also available at http://www.argo.ucsd.edu/Gould_Float_history.pdf]; Henry Charnock, "John Crossley Swallow," *Biographical Memoirs of the Fellows of the Royal Society* 43 (1997): 505–519.

77. Gould, "From Swallow floats to Argo," 530.
78. J.B. Tait, (ed.), *The Iceland-Faroe Ridge International "Overflow" Expedition, May-June, 1960: An Investigation of Cold, Deep Water Overspill into the North-Eastern Atlantic Ocean*, ICES Report 157 [available at: http://ocean.ices.dk/Project/OV60/RPV157.pdf, last accessed 10/09/2013].
79. Helen M. Rozwadowski, *The Sea Knows no Boundaries: A Century of Marine Science under ICES,* (Seattle: University of Washington Press, 2004): 133–4.
80. Hamblin, *Oceanographers and the Cold War*, 233.
81. Ibid, 232–4.
82. Minutes of Meeting, 3 September 1962, Working Group on Oceanography, CAB 124/2170, TNA (London).
83. Ryurik A. Ketov, "The Cuban Missile Crisis as Seen Through a Periscope," *The Journal of Strategic Studies* 28:2 (2005): 217–231; Geoff Till, "Holding the Bridge in Troubled Times: The Cold War and the Navies of Europe," *The Journal of Strategic Studies* 28:2 (2005): 309–337.
84. Norman Friedman, *Seapower as Strategy: Navies and National Interests* (Annapolis: Naval Institute Press, 2001): 58.
85. Head of Home NAVAL, Note on UK Participation in SOSUS, 28 December 1967, DEFE 69/526, TNA (London).
86. Visit of Team to naval facility, Lewis, Delaware, and Norfolk SOSUS evaluation centre, June 1967, DEFE 69/526, TNA (London).
87. Oceanographic Research Committee Position Paper, UK Participation in the SOSUS System, draft by DUSW (N), undated, DEFE 69/526, TNA (London).
88. Head of Home NAVAL, Note on UK Participation in SOSUS, 28 December 1967, DEFE 69/526, TNA (London).
89. ORC Position Paper, UK Participation in the SOSUS System, draft by DUSW (N), undated, DEFE 69/526, TNA (London).
90. Ibid.
91. AUWE Working Party Report, 'Research and Development Implications' (Annex 8), September 1967, in UK Participation in the SOSUS System, DEFE 69/526, TNA (London).

92. AUWE Working Party Report, 'Political Considerations' (Annex 4), September 1967, in UK Participation in the SOSUS System, DEFE 69/526, TNA (London).
93. AUWE Working Party Report, 'Financial Considerations' (Annex 3), September 1967, in UK Participation in the SOSUS System, DEFE 69/526, TNA (London).
94. Ibid.
95. For a more detailed political history of Brawdy, see Charlie Whitham, "Bargaining over Brawdy, Negotiating the American Military Presence in Wales, 1971," in *Military Bases: Historical Perspectives, Contemporary Challenges,* ed. Luís Rodrigues and Sergiy Glebov (Amsterdam: IOS Press, 2009):40–55.
96. Minute of the Minister for Defence, Lord Carrington, for the Prime Minister, 28 December 1970, PREM 15/291, TNA (London).
97. Anthony S. Laughton, "The Future of Oceanographic Research in the Light of the UN Convention," in *The UN Convention on the Law of the Sea: Impact and Implementation,* ed. Edward Duncan Brown and Robin Rolf Churchill (Honolulu: University of Hawaii, 1987): 388.
98. On the episode see Mitcehll B. Lerner, *The Pueblo Incident. A Spy Ship and the Failure of American Foreign Policy* (Lawrence: University Press of Kansas, 2002); Frank Baldwin, "Patrolling the Empire: Reflections on the USS Pueblo," *Bulletin of Concerned Asian Scholars,* (Summer 1972): 54–75.
99. Instructions to ships 'Operation CAIRNDUFF', Shadowing of Russian ELINT Trawlers, 1962, ADM 1/28093, TNA (London).
100. John Krige, *American Hegemony and the Postwar Reconstruction of Science in Europe* (Cambridge, MA: MIT Press, 2006): 4–14.
101. For more see Simone Turchetti, "Sword, Shield and Buoys: A History of the NATO Sub-Committee on Oceanographic Research, 1959–1973," *Centaurus* 54:3 (2012): 215.
102. For a history of the fishery dispute see Carmel Finley, *All the Fish in the Sea: Maximum Sustainable Yield and the Failure of Fisheries Management* (Chicago: The University of Chicago Press, 2011).

References

Aldrich, Richard. "British Intelligence and the Anglo-American 'Special Relationship' During the Cold War." *Review of International Studies* 24 (1998): 331–351.

Baldwin, Frank. "Patrolling the Empire: Reflections on the USS Pueblo." *Bulletin of Concerned Asian Scholars* (Summer 1972): 54–75.

Ball, Desmond. "Nuclear War at Sea." *International Security* 10 (1985): 3–31.

Black, Jeremey. *The British Seaborne Empire*. London: Yale University Press, 2004.
Bourgoin, J. "Henri Lacombe (33), 1913–2000. In Memoriam." *La Jaune et la Rouge* (February 2001).
Burrows, William E. *Deep Black: Space Espionage and National Security*. New York: Random House, 1988.
Calder, Kent E. *Embattled Garrisons: Comparative Base Politics and American Globalism*. Princeton: Princeton University Press, 2010.
Camprubí, Lino, Sam Robinson. "A Gateway to Ocean Circulation: Surveillance and Sovereignty at Gibraltar." *Historical Studies in the Natural Sciences* 46 (2016): 429–459.
Chafer, Tony. *The End of Empire in French West Africa: France's Successful Decolonization?* Oxford: Berg, 2002.
Charnock, Henry. "John Crossley Swallow." *Biographical Memoirs of the Fellows of the Royal Society* 43 (1997): 505–519.
Connelly, Matthew. "Rethinking the Cold War and Decolonization: The Grand Strategy of the Algerian War for Independence." *International Journal of the Middle East* 33 (2001): 221–245.
Craven, John. *The Silent War: The Cold War Battle Beneath the Sea*. New York: Touchstone, 2002.
Critchley, Harriet. "Polar Deployment of Soviet Submarines." *International Journal* 39 (1984): 828–865.
Crossman, Richard. *The Diaries of a Cabinet Minister: Volume Two*. London: Hamish Hamilton & Jonathan Cape, 1976.
Dennis, Michael A. "Earthly Matters: On the Cold War and the Earth Sciences." *Social Studies of Science* 33 (2003): 809–819.
Dobson, Alan P. *Anglo-American Relations in the Twentieth Century: Of Friendship, Conflict and the Rise and Decline of Superpowers*. London: Routledge, 1995.
Dockrill, M. *British Defence Since 1945*. Oxford: Wiley-Blackwell, 1988.
Ellingsen, Gunnar. "Varme havstrømmer og kald krig: 'Bergensstrømmåleren' og vitenskapen om havstrømmer fra 1870-årene til 1960-åreme [Warm Ocean Currents and the Cold War: The Bergen Current Meter and the Science of Ocean Currents from the 1870s to the 1960s]." PhD diss., Universitetet I Bergen, 2012.
Engel, Jeffrey A. *Cold War at 30,000 Feet: The Anglo-American Fight for Aviation Supremacy*. Cambridge, MA: Harvard University Press, 2007.
Finley, Carmel. *All the Fish in the Sea: Maximum Sustainable Yield and the Failure of Fisheries Management*. Chicago: The University of Chicago Press, 2011.
Friedman, Norman. *Seapower as Strategy: Navies and National Interests*. Annapolis: Naval Institute Press, 2001.
Gaddis, John Lewis. *Strategies of Containment: A Critical Appraisal of American National Security Policy During the Cold War*. Oxford: Oxford University Press, 2005.

Galison, Peter. "The Many Faces of Big Science." In *Big Science: The Growth of Large Scale Research*. Edited by Peter Galison and Bruce Hevly. Stanford: Stanford University Press, 1992.

Gold, Peter. *A Stone in Spain's Shoe: The Search for a Solution to the Problem of Gibraltar.* Liverpool: Liverpool University Press, 1994.

Gould, W. John. "From Swallow Floats to Argo – The Development of Neutrally Buoyant Floats." *Deep-Sea Research* 52 (2005): 529–543.

Grove, Eric. *Vanguard to Trident: British Naval Policy Since World War II.* Annapolis: Naval Institute Press, 1987.

Hahn, Peter. *The United States, Great Britain, Egypt, 1945–1956: Strategy and Diplomacy in the Early Cold War.* Chapel Hill: University of North Carolina Press, 1991.

Hamblin, Jacob D. *Oceanographers and the Cold War: Disciples of Marine Science.* Seattle: University of Washington Press, 2005.

Harkavy, Robert E. *Bases Abroad: The Global Foreign Military Presence.* Stockholm: SIPRI, 1989.

Herran, Nestor, Xavier Roqué. "An Autarkic Science: Physics, Culture and Power in Franco's Spain." *Historical Studies in the Natural Sciences* 43 (2013): 202–235.

Huchthausen, Peter, Alexandre Sheldon-Duplaix. *Hide and Seek: The Untold Story of Cold War Naval Espionage.* London: John Wiley, 2009.

Ketov, Ryurik A. "The Cuban Missile Crisis as Seen Through a Periscope." *The Journal of Strategic Studies* 28 (2005): 217–231.

Krige, John. *American Hegemony and the Postwar Reconstruction of Science in Europe.* Cambridge, MA: MIT Press, 2006.

Laughton, Anthony S. "The Future of Oceanographic Research in the Light of the UN Convention." In *The UN Convention on the Law of the Sea: Impact and Implementation.* Edited by Edward Duncan Brown and Robin Rolf Churchill. Honolulu: University of Hawaii, 1987.

Leffler, Melvyn P. *A Preponderance of Power: National Security, the Truman Administrations, and the Cold War.* Stanford: Stanford University Press, 1992.

Lerner, Mitchell B. *The Pueblo Incident: A Spy Ship and the Failure of American Foreign Policy.* Lawrence: University Press of Kansas, 2002.

Macrakis, Kirstie. "Technophilic Hubris and Espionage Styles During the Cold War." *Isis* 101 (2010): 378–385.

Mastney, Vojtech. "The 1963 Nuclear Test Ban Treaty: A Missed Opportunity for Détente?" *Journal of Cold War Studies* 10 (2008): 3–25.

Mort, Maggie. *Building the Trident Network: A Study of the Enrolment of People, Knowledge and Machines.* Cambridge, MA: MIT Press, 2002.

Nalebuff, Barry. "Brinkmanship and Nuclear Deterrence: The Neutrality of Escalation." *Conflict Management and Peace Science* 9 (1986): 19–23.

Nechtman, Tillman. "'…for It Was Founded Upon a Rock': Gibraltar and the Purposes of Empire in the Mid-Nineteenth Century." *The Journal of Imperial and Commonwealth History* 39 (2011): 749–770.

Nolfo, Ennio Di. "The Cold War and the Transformation of the Mediterranean." In *The Cambridge History of the Cold War.* Edited by Melvyn P. Leffler and Odd Arne Westad. Cambridge: Cambridge University Press, 2010.

Oreskes, Naomi. "A Context of Motivation: US Navy Oceanographic Research and the Discovery of Sea-Floor Hydrothermal Vents." *Social Studies of Science* 33 (2003): 697–742.

Polmar, Norman, K.J. Moore. *Cold War Submarines: The Design and Construction of US and Soviet Submarines.* Washington: Potomac Books, 2004.

Reed, W. Craig. *Red November: Inside the Secret US-Soviet Submarine War.* London: Harper Collins, 2010.

Richelson, Jeffrey T., Desmond Ball. *The Ties that Bind: Intelligence Cooperations Between the UKUSA Countries- the United Kingdom, the United States of America, Canada, Australia and New Zealand.* London: Unwin Hyman, 1990.

Robinson, Sam, Lino Camprubí. "Ocean Science, Gibraltar and Geopolitics – Then and Now." Science Blogs – Political Science (Blog). *The Guardian.* http://www.theguardian.com/science/political-science/2013/oct/04/ocean-science-gibraltar-geopolitics/.

Rozwadowski, Helen M. *The Sea Knows No Boundaries: A Century of Marine Science Under ICES.* Seattle: University of Washington Press, 2004.

Sangmuah, Egya N. "Eisenhower and Containment in North Africa, 1956–60." *Middle East Journal* 44 (1990): 76–91.

Seaborg, Glenn T. *Kennedy, Khrushchev and the Test Ban.* Berkeley: University of California Press, 1981.

Snyder, W.P. *The Politics of British Defence Policy.* Athens, OH: Ohio University Press, 1964.

Soloman, James B. *The Multilateral Force: America's Nuclear Solution for NATO (1960–1965).* Annapolis: US Naval Academy, 1999.

Sontag, Sherry, Chistopher Drew. *Blind Man's Bluff: The Untold Story of American Submarine Espionage.* London: Harper Collins, 1998.

Spinardi, Graham. *From Polaris to Trident: The Development of US Fleet Ballistic Missile Technology.* Cambridge: Cambridge University Press, 1994.

Till, Geoff. "Holding the Bridge in Troubled Times: The Cold War and the Navies of Europe." *The Journal of Strategic Studies* 28 (2005): 309–337.

Turchetti, Simone. "Sword, Shield and Buoys: A History of the NATO Sub-Committee on Oceanographic Research 1959–1973." *Centaurus* 54 (2012): 205–231.

Twigge, Stephen, Edward Hampshire, Graham Macklin. *British Intelligence Secrets, Spies and Sources.* London: The National Archives, 2009.

Watt, Donald C. *Succeeding John Bull: America in Britain's Place 1900–1975.* Cambridge: Cambridge University Press, 1984.

Weir, Gary E. *An Ocean in Common. American Naval Officers, Scientists, and the Ocean Environment.* College Station, TX: Texas A&M University Press, 2001.

Whitham, Charlie. "Bargaining Over Brawdy, Negotiating the American Military Presence in Wales, 1971." In *Military Bases: Historical Perspectives, Contemporary Challenges.* Edited by Luís Rodrigues and Sergiy Glebov. Amsterdam: IOS Press, 2009.

Wright, Tom. "Aircraft Carriers and Submarines: Naval R&D in Britain in the Mid-Cold War." In *Cold War Hot Science: Applied Research in Britain's Defence Laboratories 1945–1990.* Edited by Robert Bud and Phillip Gummett. London: Science Museum, 1999.

CHAPTER 6

Militant Oceanographers: Behind Britain's "Technocratic" Moment, 1958–64

"The sea is whose oyster?" asked the headline of an article in *The Statist* which opened,

> The National Institute of Oceanography had its first open day last week. And not before time; the Institute has been in existence for 16 years but has only just discovered that a very important way of influencing the Treasury is by generating publicity. The Institute needs more money and, considering the quality of work performed on its present shoestring, it deserves to get it.[1]

For *The Statist* oceanography provided a 'text book example of the problems of investment in fundamental science', mainly that investment was not comparable to that of both Britain's allies and enemies, and worse, of the other Commonwealth nations.[2] Whereas justifying oceanography economically was fraught with difficulties, the case for 'keeping up with the international Joneses' was, the article maintained, 'very strong'.[3] This was seen as an acceptable form of justification because now the Western alliance had come to rely on research that was shared internationally on a mutual basis, as shown in the previous chapter. However, this system only functioned as long as others in the British government valued research carried out in Britain. *The Statist* blamed both the NIO managers for their perceived lack of political awareness and the Treasury for their 'miscomprehension' of the importance of maintaining Britain's reputation as a leading international scientific power. This need for new sponsors and

S. A. Robinson, *Ocean Science and the British Cold War State*,
Palgrave Studies in the History of Science and Technology,
https://doi.org/10.1007/978-3-319-73096-7_6

greater public exposure stemmed from fundamental changes in the approach to science policy, governance, and patronage within the British state during the premiership of Harold Macmillan (1957–63).

The relationship between government, the military, and the NIO changed during the early 1960s as new funding bodies were created, civil science was reorganised, and military science was separated administratively from civilian research establishments. Those strands of oceanographic research of use to surveillance and naval operations were increasingly being brought within defence research establishments, and no longer left in the hands of civilian scientists. Even more worryingly the transfer of the NIO from the control of the Ministry of Defence (MoD) to the newly formed Natural Environment Research Council (NERC), a civilian body, significantly reduced the opportunities for collaboration between the NIO and the Admiralty. Despite substantial support from the military to bolster oceanographic studies in Britain since the war, the MoD consented to this displacement of an influential international centre for ocean science from a military to a civilian authority, with little protest.

These changes resulted from the pursuit of a new science policy in Britain, which questioned the relationship between science funding and the defence budget. These were significant for the NIO and oceanography in general. The notion of what oceanography could offer changed rapidly during the early 1960s, as civilian applications began to rival existing military ones. In an effort to counter the swift changes in patronage, oceanographers and science communicators advocated the potential of oceanography as a peaceful science which offered more possibilities than the rapidly expanding field of space research. One such example of this lobbying occurred in 1963 when Frederick Brundrett claimed in a lecture, 'A World which spends more in a single year on research into space than it has done since the beginning of the century on research into the sea is mad.'[4] Within this changing policy landscape and growing popular interest in the new Big Science of the oceans and space, the Royal Society administrators came once again to the assistance of the British oceanographic community.

These transitions also echoed important debates on science policy then occurring in Britain. The five years leading up to the 1964 general election have been described by various historians and political scientists as marking the beginning of Britain's brief technocratic moment.[5] In 1959 the British intellectual and civil servant C.P. Snow famously lectured on the "two cultures" present in the British administration, calling for greater awareness of the potential of science for British society.[6] Five years later a Ministry of Technology was established, and "modernity" was seemingly placed at the forefront of political and cultural discourses.[7] As David Edgerton argues:

> The technocratic moment was *ideologically* much more significant than has been recognized, and much less significant in policy terms: it defined a whole mode of thinking about the British élite and state, and British science, technology and industry in particular.[8]

It is the central argument of this chapter that not only was this moment ideologically significant but it did have an impact on ocean science policy. Oceanography was one of a small group of sciences that had grown out of wartime work, but never full separated from the military machinery of government that allowed the discipline to expand after 1945. The policy significance of Britain's technocratic moment only emerges when considering the experience of smaller scientific disciplines such as the ocean sciences.

Histories of British science and technology looking at this period have tended to focus their attention on what may be termed "prestige" projects. Thus, histories of specific projects that had a great deal of state support during the period, including the building of the motorway network, the supersonic airliner project (Concorde), the construction of second-generation Cold War military technologies (nuclear submarines such as Polaris, and the Harrier Jump Jet), have all been written and re-written. Other proposed projects never made it beyond 1964, when several projects were cancelled by the new Labour government headed by Harold Wilson. Examples include the CVA065 (super-aircraft carrier), TSR2 (tactical jet fighter-bomber), and the notion of an independent British space programme.[9] This has resulted in the historiography of scientific–political interaction during the period being scattered amongst various project-specific histories.

Behind all of these projects, which have been said to encapsulate the "technocratic" moment, there were other sciences that struggled to capture political support. However, if historians only focus on grand projects we get a unilateral history of success and failure, with no middle ground; some sciences certainly benefited from greater access to funds during this period, but others struggled. Oceanography provides one such example. Whilst to the outside world British oceanography seemed to be flourishing—with a new vessel and a prestige project in the Indian Ocean—the reality back home was, as this chapter shows, really very different. At a time when domestically, as seen in the previous chapter, some scientists sought to undermine Deacon because of his positioning in the international arena (especially through NATO), he struggled to secure both an increase in funding and a commitment to longer-term funding for the NIO. This chapter charts the struggle that took place behind the technocratic moment, as civil servants and especially the Treasury attempted to allocate funds to a wide variety of military and civil science and technology commitments. The technocratic "bubble" began to burst.

In this chapter I argue that during the early 1960s Deacon and his allies promoted a public-wide campaign to raise awareness of the potential of oceanography in an attempt to further expand the policy network set up in the 1950s, and bring together the advantages for the NIO to be derived from military funding with the popularity and further prestige of a civilian programme. However, this public campaigning did not succeed in creating the circumstances for a further expansion of Deacon's network. If anything, it anticipated its crumbling. By the mid-1960s major changes in the NIO administration relinquished vital ties with military patrons, marking an important transition for the NIO from military to civilian oceanography.

In raising awareness for oceanography Deacon sought to acquire new sources of funding for ocean science by campaigning through three channels: the popular press, the Royal Society, and the civil service. Before 1963 these three channels, when presented with a unified approach, had often allowed for the continued expansion of NIO programmes. Following exchanges in 1960 the NIO secured a new research vessel, a commitment to an increase in scientific staff, and the refurbishment of the buildings at the NIO. Subsequently, those operating outside the auspices of the NIO felt aggrieved about their limited access to "national" resources, and the consensus amongst oceanographers broke down. The 1963 Report of the Committee of Enquiry into the Organisation of Civil Science in the United Kingdom (commonly known as the Trend Report from the name of the committee's chair, Burke Trend) had a definite impact on oceanographic science in Britain, although the overall impact of the report may have been limited for science more generally.[10]

For oceanography, the definition of military and civilian as separate spheres was a cause of concern for Deacon who felt that oceanographers needed defence-sized budgets so that they could carry out "big science" projects comparable to those of space research. The result was that in order to place science in the hands of scientists, the actual views of the scientists themselves were less relevant than the policy itself.

Promoting the Gospel of Marine Science

In August 1959, the close confidant and backer of George Deacon, Sir Frederick Brundrett, stated in *The Sunday Times* that 'Britain's future as a maritime nation is in serious jeopardy because of lack of research in the marine sciences'.[11] Brundrett had risen through the ranks to become the chairman of the Defence Research Policy Committee and Chief Scientific Adviser to the Ministry of Defence, and he also represented Britain internationally in forums such as the NATO Defence Research Group (a group which oversaw the work of the NATO Science Committee and its various sub-committees' programmes). Brundrett was not the only one to ring alarm bells: an editorial in *New Scientist* commented a week later, 'Russia's marine science programme is almost comparable with her space research'.[12] Concurrently the US Congress was debating increasing funding for oceanography from US$8 million to US$28 million, and several smaller nations had considerably enlarged their oceanography programmes. British efforts had, by comparison, been less spectacular.[13] Brundrett's piece in *The Sunday Times* prompted *New Scientist* to commission a series of articles reporting recent developments in oceanography.

Brundrett and Deacon collaborated to disclose the funding issues around oceanographic work in an effort to find new patrons. They produced an article for leading industry-specific periodical *Shipping* detailing what they saw as a lack of research funding for ocean sciences. Deacon trusted Brundrett to support his oceanographic interests. Their relationship had become increasingly informal ever since they had worked together to persuade the Admiralty of the benefits of expanding oceanographic research in 1945.[14] Brundrett believed it was important to 'improve and extend our research in this country in the science of the sea generally'.[15] He sought to promote discussion beyond the scientific, military, and governmental communities that were primarily concerned with oceanography (its applications, funding, and growth) into new and more commercial avenues such as the shipping industry.[16] By 1959, the NIO had diversified, contributing to a range of outside agencies providing applications for their research. For instance, publishing Brundrett's article in *Shipping* was justified by Deacon as a result of the work the NIO had been carrying out with the British Shipbuilding Research Association and the Ship Division of the

National Physical Laboratory studying the motion of ships at sea.[17] Brundrett's article for *Shipping* sought to compare the ocean and outer space in the context of the developing scientific Cold War. His objectives were clear:

> Which would you think more important to the future of the world – research in space or research in the sea? Whatever your answer to that question, I think you would be surprised to learn that the world is now spending more in one year on research in space than it has spent on sea research this century.[18]

Brundrett was not the only Briton asking these questions. As shown by Helen Rozwadowski, at this time the renowned author Arthur C. Clarke was arguing that both the ocean depths and space represented a twentieth-century "frontier" for mankind.[19] This duality of oceans and space was reflected in Clarke's work: in 1960 he wrote a non-fiction work entitled *Challenge of the Sea* and also published the novel *A Fall of Moondust,* set on the moon.[20] Internationally, French ocean explorer Jacques Cousteau and his team of divers had brought oceanography, albeit in the format "popular science meets wonders of the deep", to film screens, and featured on the front cover of *Time* in 1960.[21] As a propagandist for ocean science, Cousteau united research and adventure in the deep sea, along with the evolving technological prowess of undersea exploration.[22] Evidently the situation in France was very different from that in Britain. In France, the prestige and popular appeal of Cousteau's enterprises helped Henri Lacombe and other French oceanographers to find the resources necessary to promote sizeable projects (within and outside NATO) and continue to collaborate with naval authorities. This was a moment in which both space and the ocean seemed to provide the geographical context for the new frontier for scientific exploration.

Whilst the use of a frontier analogy allowed a literary and cultural interaction with both the oceans and space, scientists in the United States had used the same metaphor to invoke a new age of marine science as early as 1953 at a meeting of the American Association for the Advancement of Science, during a special session on "The Sea Frontier".[23] This session included various papers on the potential of geological and biological resources of the oceans. In 1963 Cousteau proposed to the World Congress on Underwater Activities (a diving rather than scientific conference) that humanity had the potential to evolve into a being that could breathe underwater—*homo aquaticus*—informing the world that NASA was working on gills that allowed humans to breathe.[24] Whilst Cousteau's and Clarke's works demonstrated that science fiction dreams drove ocean science as much as space science, those attempting to direct investment in ocean science attempted to draw clearer practical benefits beyond the military. Brundrett was arguing that it was more likely that the sea offered greater riches than the vacuum of space.

Recent historiography has attempted to tie the history of US Cold War science to that of the European experience, but perhaps the United States should be seen as the exception rather than the rule.[25] The United States could afford to conduct space and ocean research simultaneously. In contrast, Brundrett would have been all too aware that recent British experience of science funding, combined with a drive to cut government spending in general, made comparable investment impossible. Between 1957 and 1963, when the Conservative Harold Macmillan was Prime Minister, the government's agenda was to judge scientific projects based on economic benefit within limited fiscal reserves, and thus substantially reduce the other programmes' budgets accordingly. Brundrett spoke of importance, but this was not a philosophical question about scientific merit, it was a more difficult question about value, expenditure, and reward. In Brundrett's article, reward was viewed differently from the utopian images invoked by Clarke. The latter's novel *The Deep Range* had described vast plankton farms and herds of whales, upon which one character reflected, 'until the oceans froze, he [man] would never be hungry again'; in Britain, defence and industry took precedence.[26]

In his article Brundrett used six sub-headings: 'Defence', 'Ship Design', 'Navigation', 'Coastal and Harbour Engineering', 'The Effect of Oceans on Climate', and 'Use of the Sea for Waste Disposal'. Brundrett fell back on oceanographers' arguments about the submarine menace of the previous two World Wars, but he also looked forward to the new threat from the Soviet Union, which he claimed had the largest fleet 'the world has ever known'.[27] The article discussed the advent of the nuclear submarine, claiming that seaborne acoustic equipment had reached the end of its development potential, and that the only hope for improving this technology lay in a better understanding of the 'characteristics of the sea as a medium'.[28] However, Brundrett was just about to leave his position as Chief Scientific Adviser and Chairman of the Defence Research Policy Committee at the Ministry of Defence and as a defence research director at NATO. As he was soon to take up the position of Scientific Commissioner to the Civil Service, where he would be unable to inform the patronage of oceanography from the corridors of power, Deacon lost a key ally in the Admiralty and the Ministry of Defence.[29] Concurrent with Brundrett's campaigning in the press, Deacon wrote a piece intended for dissemination to a wider public advocating the expansion of the marine sciences. Deacon also composed an introduction to a special series of articles for *New Scientist* on oceanography, commissioned following Brundrett's comments in *The Sunday Times*.[30] In addition to these campaigns to raise the profile of oceanography amongst popular newspapers, magazines, and trade periodicals, they sought to mobilise the Royal Society.

A public campaign to raise awareness of ocean science could not inform new policies directly; changing policy required wider institutional and establishment support. Alongside writing for the popular press, Deacon and Brundrett also influenced the activities of the Royal Society in order to further shape new policies supporting ocean science. The Society had to tread a fine line between its lobbying for government patronage of science and its independence as a scientific body. As the founding of the NIO had shown, the Society often played a dual role, representing British science in the international arena, whilst simultaneously attempting to shape government policy for science through the fellows of the Society who sat on various government advisory committees. In this dual role some of the Royal Society administrators, including its new secretary, wished to favour the growth of oceanography and ensure a level of patronage at home that would keep British oceanographers in influential positions in the international sphere.[31]

This support materialised in the establishment of the Royal Society's British National Committee for Oceanographic Research (BNCOR) in 1959.[32] This body now took responsibility for advising on oceanography funding policies alongside the National Oceanographic Council (NOC) which had been brought into existence by the same Royal Charter that had established the NIO in 1949.[33] But unlike the NOC, which was chaired by a representative of the Admiralty (usually the CRNSS), the BNCOR was an advisory body comprising only scientists. The Royal Society's Assistant Secretary (who later became the Executive Secretary), David Christie (D.C.) Martin, agreed to support Deacon's appointment as chair of the new committee in an effort to strengthen his positon.[34] Martin had agreed to form this committee alongside several others because of the Royal Society's position as the UK representative to the International Council of Scientific Unions (ICSU), the world-wide organisation representing scientific academies.[35] In order to advise on specific disciplines which had been united in large organisations under the ICSU aegis, Martin felt that a new advisory body was needed.

This was to some extent a reaction to what was happening at the international level. Following the IGY several governance changes had occurred at ICSU; its Scientific Committee for Oceanographic Research (SCOR) had replaced the International Joint Commission on Oceanography (JOC). The establishment of SCOR was controversial. Propagandised as an attempt to combine biological and geophysical oceanography, it reflected Deacon's vision of combining biological, physical, and chemical approaches to the study of the oceans so as to—ultimately—prioritise the physical. As Hamblin has shown, several international oceanographers who were aware of his hidden agenda blamed Deacon for the disbanding of JOC.[36] SCOR brought together a group of international oceanographers keen to run the first large-scale international expedition free from military/governmental interference, unlike the IGY and the newly formed NATO sub-committee on oceanographic research. In these arenas physical oceanographers sidelined their colleagues devoted to fishery studies and marine biology. Biological oceanographers blamed Deacon for what they saw a political move to promote his brand of ocean science.

The first major proposal from SCOR was to follow the IGY with an expedition to the Indian Ocean (IIOE), relatively under-explored by physical oceanographers and at this time (before the Vietnam conflict) a strategically non-volatile ocean space. So while NATO was now dominated by physical oceanographers interested in the exploration of areas of interest to the Western alliance, the physical oceanographers in SCOR set the locations of international expeditions to avoid these spaces, thereby formulating a skewed strategic vision for oceanography outside areas of military interest. These shifts in governance further reduced the visibility of biologists and fisheries scientists at the international level. But this evident shift towards physical oceanography over biological approaches was not readily apparent to those outside the discipline and seemed to raise no concerns at the Royal Society, which had established the BNCOR in an effort to coordinate the activities promoted internationally by the ICSU-sponsored SCOR. As the IIOE would include fisheries research, but would notably not include any objective to develop local fisheries beyond the scope of the expedition, it appeared that the inclusion of fisheries research was a token gesture.

As the BNCOR was a body that fulfilled part of the Royal Society's international obligations to ICSU, D.C. Martin took a particularly keen interest from the outset, recognising that exchanges with Deacon would prove mutually beneficial. Martin's involvement gave Deacon access to the secretary's network of civil servants and offered him another route to gain influence in Whitehall. Deacon was a leading SCOR member and operated freely in the international oceanographic community. He could therefore introduce Martin to the mechanisms of international collaboration. Each benefitted from the other's expertise. Martin had always operated as a go-between for the Royal Society and Whitehall, and as shown by Jeff Hughes, had been undertaking this role since the late 1940s.[37] Furthermore as shown by Jennifer Goodare, the IGY represented a valuable opportunity for the Royal Society to mobilise its fellows and organise its activities under the national IGY committees in order to show how the Society could successfully contribute to the British government's science policy.[38]

Throughout 1960 Martin became interested in the activities of British and international oceanography.[39] Writing to Deacon on 23 January 1960, Martin acknowledged that he was no expert in the diplomacy of international oceanography, remarking 'In what follows you may think I am talking nonsense and if I didn't know you so well I shouldn't write at all'.[40] It was no coincidence that Martin's interest in oceanography emerged at this time, as 1960 marked the tercentenary of the Royal Society. This publicly highlighted the work of the Royal Society, whose role had been steadily declining nationally as government machinery for science became more comprehensive, whilst expanding internationally.[41] At the Tercentenary Banquet held on 26 July 1960, the Prime Minister Harold Macmillan stated: 'it is part of your scientific function, your great humanist function, to continue that long tradition of co-operation between nations and the free exchange of knowledge.'[42] The Royal Society aspired to control the management of British science whilst emphasising its position 'at the heart of both the global scientific community and the British Establishment'.[43] To this end the outgoing President of the Royal Society, Cyril Hinshelwood, had begun to consider how the Society might retain a role, given the expansion of scientific advisory bodies within the government machine.[44] During this period the Macmillan government was carefully considering future science policy and its potential implications. The Royal Society wanted to be a part of these discussions; the reports of committees within the Royal Society such as the BNCOR constituted early attempts at push-back against the government science policy establishment.

The mutually beneficial relation between Martin and Deacon was intended to bear fruit not just in terms of prestige and national recognition but also in terms of patronage. Martin represented for Deacon another connection allowing him to strengthen his role as node at the centre of policy networks. In May 1960, the NIO director prepared a memorandum on the forthcoming International Indian Ocean Expedition (IIOE) that the SCOR had discussed. The memorandum was to add weight to an approach by the NOC to the Treasury to ask for money to support the proposed expedition.[45] Deacon's BNCOR memorandum had been sent to the council of the Royal Society. Martin now took the opportunity to suggest that a discussion meeting with an attached conference should be arranged so that the 'whole question of oceanography' in its widest sense could be considered.[46] The BNCOR memorandum essentially invited the government to accept in principle the need for increasing national research in oceanography on a long-term basis, even doubling expenditure in the next five years if necessary. It also called on the government not to restrict the recruitment of more staff into government laboratories and to encourage the interest of the universities in the marine sciences. It finally invited the Admiralty to consider the commissioning of an additional research ship for university scientists.[47]

The memorandum concluded that international expansion of oceanographic research was likely to take place and that Britain needed to be a part of this growth. British scientists would therefore be in a position to interpret and utilise this expanding body of knowledge to solve the 'considerable maritime needs of the UK'.[48] The timing of the meeting, a few months before the Treasury/Admiralty inspection of the NIO, gave Deacon a political advantage,[49] allowing him to use the conference's concluding remarks to air his views. Deacon assumed he would receive Treasury backing because he had received that from Martin and the Royal Society. However, Treasury officials took the view that more support for oceanography would not be decided upon on the basis of international prestige or even the support of Britain's oldest and most prestigious scientific society. In the confines of the Royal Society Deacon could advocate whatever he liked and be praised by Martin for it, but actual change required direct formal negotiation with government officials.

'We Can Go on If We Are Not Fitted into a Treasury Straitjacket'[50]

Throughout 1949–65, the NIO budget was the subject of a continual battle between Deacon and the NOC on one side, and the Admiralty and the Treasury on the other. However, agreement had been reached for the five years from 1957–58 to 1961–62, with the grant-in-aid to the NIO set at £175,000 per annum. This figure had been set during the IGY, when oceanography sat briefly in the limelight alongside space science. The NIO inspection carried out in 1960 can be seen as the beginning of the Macmillan government's review of science policy vis-à-vis defence policy. Deacon saw this not as an opportunity but more as an intrusion. In a letter to the CRNSS he described the inspection as 'an intolerable burden', which had given him no help whatsoever.[51] He was sure of the contribution that the NIO made and his own role within it. 'I don't want to blow my own trumpet, but I believe this laboratory is something out of the ordinary by Civil Service standards, and it would suffer if these were too rigidly applied. Look what has been done in co-operation with Woods Hole and the US NSF [National Science Foundation] – a very un-treasury-like operation that has written a new chapter in the study of ocean circulation.'[52] Here he was alluding to the recently completed IGY study of the Gulf Stream which had been undertaken alongside WHOI. Deacon's emphasis on joint Anglo-American research should not be regarded as acceptance of any kind of subservient or even equal relationship between the nations. He argued that 'five Englishmen can go to the USA and more than hold the reputation of the UK for ability in theoretical and practical research against 50 Americans spending 50 times as much'. In order to understand Deacon's views at this time it is important to contextualise the funding for ocean science within broader changes in defence expenditure, as it was from the defence budget that deep-sea oceanography was funded.

When Harold Macmillan became Prime Minister in January 1957 he had stated that 'no vested interest, however strong, and no traditions, however good' would halt a reassessment of defence policy.[53] Macmillan appointed Duncan Sandys as Minister of Defence and instructed him to substantially reduce defence expenditure. Sandys' methods and approach were disliked by the Chiefs of Staff.[54] His 1957 Defence White Paper proposed savage cuts to the budget to bring spending into line with that of Britain's post-war NATO allies. Despite demanding cuts of £180 million from the defence budget, the White Paper did little except halt incremental growth in funding at the NIO. The White Paper had expressly focused on the nuclear deterrent as a justification for reducing conventional forces.

By 1960 several crises forced a reassessment of the strategic ideology proposed by the 1957 White Paper. Macmillan was aware that Britain had to accept it had a crumbling empire and was overstretched.[55] This forced the government to call up reservists and retain National Service personnel beyond the intended official termination of the programme. A new Defence White Paper in 1960 revised Sandys' nuclear deterrence model that had prioritised missiles over conventional forces. The 1960 White Paper stated that 'because of the need to meet local emergencies which could develop into a major conflict, conventionally armed forces are a necessary complement to nuclear armaments. In short, the deterrent to aggression must be comprehensive.'[56]

Following this continued uncertainty in defence strategy and government spending the Civil Lord of the Admiralty, Ian Orr-Ewing MP, asked for a private report into the activities of the NIO.[57] The note attached to the report indicates that the review had been suggested by Orr-Ewing who had been looking to rationalise the means of financing the Institute. The summary of the report concluded: 'there is probably more direct benefit to the Navy from the research work of the NIO than from the work of the observatories, whose costs we also bear on Navy Votes, and at a time when defence demands on the NIO show signs of growing, it would seem imprudent to do anything which might diminish our present influence.'[58] The report reveals much about the methods and motivations of Brundrett, who had first raised the possibility of making greater use of the NIO for anti-submarine warfare work to the Minister of Defence the previous year.[59] The author of the report agreed with Brundrett, stating that although the work of the Institute was consistent with several naval interests, the greatest promise lay in research into undersea warfare, and highlighted potential NIO involvement with a recently founded NATO ASW Research Centre at La Spezia (Italy).[60] Several NIO oceanographers, including Henry Charnock, would come to play key roles in the development of this research centre. Nevertheless, there was little point in the British duplicating the US effort—otherwise the collaborative transnational direction of national research efforts, a motivation for NATO's scientific committees, would fail to bring about collective benefits.[61] Therefore, a wide breadth of research approaches was required to contribute at the same level as the USA; the challenge was to conduct this within the UK's much smaller GDP. The report, however, did not directly advocate the allocation of extra funds. In 1960, the primary issue appeared to be the lack of suitably qualified staff, something Deacon regularly highlighted but blamed on the universities.[62]

In the summer of 1960, a Treasury official approached the Civil Establishments Branch of the Admiralty to organise an inspection of the grading of staff at the NIO. During the previous five years, the Treasury had attempted to inspect the Institute in 1955 (at the Admiralty's suggestion), then in November 1956, again in 1957, and in 1958.[63] In each case Deacon had managed to persuade the potential inspectors that his scientists were too busy to be inspected at that time. Five-year cycles of funding, known as a *quinquennium* within the civil service, were the preferred way of supporting government institutions, providing the necessary stability required to keep staff and to develop research programmes. The chief of the Civil Establishments Branch contacted Deacon, who managed to persuade him that conference attendance, sea trials, and staff leave would make any inspection impractical before October. Deacon was on the defensive from the outset, suggesting that although he welcomed help with the grading of the scientific staff, he was concerned that this was an attempt by the Treasury to cut the numbers of non-scientific staff at the Institute. The chief warned Deacon that the Treasury intended to expand this into a constitutional issue if pushed, and they therefore suggested that the Treasury 'reconsider' the scope of the inspection.[64] However, it appears Deacon did little more to prepare for the coming ordeal than to protest about its rather vague terms of reference.[65]

The Admiralty/Treasury inspection of October 1960 was a much more substantial undertaking than Orr-Ewing's private report. Two representatives of the Admiralty, E.W. Pratt (Assistant Director of Physical Research) and R. A. Devereux (Civil Establishments Branch), along with an official from the Treasury, spent ten days at the NIO conducting their investigations. Alongside the official report, Pratt, the "scientist" on the panel, submitted his own supplementary note examining scientific activities of the Institute. The supplementary note sought to give the inspectors an avenue to judge the personnel required by the Institute in relation to its present and future function. Pratt was at pains to highlight the 'value of the Institute' in his note. He emphasised the NIO's reputation at home and abroad, Deacon's tolerance of individuals and their specialisms, the value of the Institute's progress in the exchange of ideas between nations, and its potential to assist Commonwealth nations with their own problems. This international outlook was praised as a positive reflection on the nation as well as the Institute. However, this positivity was somewhat balanced by advice that the Institute needed to 'weigh carefully the chances of success in the various fields of research which can be attempted with its limited resources and rate of growth and allocate rather firm priorities accordingly'.[66] This contradicted the notion that Deacon's flexibility was an institutional asset.[67]

The official report concluded that, whilst the Institute was undoubtedly 'engaged on important and necessary work and that the overall staffing was reasonable', changes in numbers and grades would be needed.[68] The controversy the report caused within the NIO was immediate. The NIO Secretary, R.G. Williams, attempted to distance himself and the NIO from the conclusions of the report. Rather than take Deacon's views into account and review the output of the Institute, the inspectors had walked around, interviewing the scientific staff whilst they worked. There had been little time to brief staff, and this approach made it difficult to argue that the inspection had been anything but thorough. In a further letter, Williams attempted to outline this: 'you have recorded various expressions of view by members of the staff which are not necessarily the official views of the Institute (or even those of the people concerned if they had reflected fully).'[69] The cause of these apparent discrepancies between unofficial and official views was the head of the biological section, Neil A. Mackintosh.

Mackintosh took the inspection as his opportunity to announce that the whale scientists at the Institute were returning to the Natural History Museum to form a Whale Research Unit. This had not been agreed with Deacon or the NOC but revealed their dissatisfaction with the way whaling scientists had been treated at the NIO. The historian of whaling science Graham D. Burnett casts the relationship between Deacon and Mackintosh in an extremely negative light, even going so far as to suggest that Deacon made accusations alleging Mackintosh was a homosexual in an attempt to reduce his influence.[70] What is certain is that Mackintosh had been demoted in 1960, and probably harboured some resentment towards his former subordinate (Deacon) who had now become his superior.[71] The movement of the cetacean scientists, as Burnett suggests, was a form of retirement plan for Mackintosh.[72] Mackintosh was reluctant to leave London when the discussions of the location of the institute took place in 1949–52 and had spent most of the last decade complaining about the location of the Discovery specimen collection (at the National History Museum) and their distance from Wormley.[73] The announcement of the imminent relocation of the whale scientists pushed the relationship between Deacon and Mackintosh to breaking point. Deacon wrote to Willis (CRNSS) 'I don't believe in the move of the whalers to London…I feel that the younger men are being sacrificed for the sake of Mackintosh. I am fairly certain that they don't want to go with him. I hope the move can be delayed a bit till things sort themselves out.'[74] Whilst Deacon was frustrated by Mackintosh's move, he was equally concerned about the potential political responses to the report.

After 1960, the country's financial difficulties did not stop increased funding for the NIO being agreed. A refinancing agreement was reached between the Development Commission and the Admiralty following the loss of dominion support for the NIO, and finally the provision of £800,000 for a new research vessel and expansion of the NIO buildings at Wormley was agreed. This was the time when most NIO work nationally (in collaboration with Admiralty research groups) and internationally (within NATO and SCOR) gave Deacon and the institute he had contributed to establish prestige and recognition. It was a moment for him to shine both as a scientific leader and a representative of British science abroad. By 1962–63, Britain had reached a crossroads—it needed to either radically reduce its global commitment or sharply increase its defence expenditure.[75] This had an impact on the financial stability of the Institute.

It would be wrong to characterise this period as one in which state funding for science was being withdrawn. Macmillan's government seemed to have little restraint when it came to funding large military or infrastructure projects and it is from this period that most notions of technocratic modernism derive, but the later 1963 Treasury inspection eventually led to cuts to the NIO budget. Ultimately the position outlined in the pre-report to the inspection—'Oceanography is so generally taken to mean ships – and ships the Admiralty'—became the prevailing reason why the NIO should be under Navy patronage.[76] This stemmed from an assertion by the Treasury that only one ministry and therefore one vote should fund governmental bodies. The Treasury premise for deciding who should pay was a simple matter of determining which ministry benefitted the most from that body; this was often the cause of much debate over the usefulness of scientific research. In order to achieve this post-war settlement Deacon had had to reach an uneasy compromise with the biologists when the NIO was first established, promising that their work would be protected under his directorship. It was on the basis of this agreement that the Admiralty and Treasury had both agreed to fund the enterprise in 1948. Yet, in the intervening period several events had contributed to tarnish Deacon's relationship with the biologists, not least the international priority being given to physical oceanography and Deacon's support of these initiatives instead of supporting the same multi-disciplinary approach he supposedly favoured domestically. With consensus breaking down the Treasury became concerned about the funding for an institution that was increasingly breaking away from its mixed studies and appeared to be prioritising defence research over biological studies.

There is a contradiction between the public face of the NIO during these years and the continuing discussions between scientists (in government employment) and low-level civil servants over funding for the NIO activities. In a letter to Sir William Cook (Chief Scientist at the Atomic Energy Authority) Deacon confided that when in 1961–62 the government gave £95,000 less than the £360,000 the NIO had requested, there had been little protest on the part of the Institute. Deacon claimed that the reason for this lack of protest stemmed from having been given a 'new ship and the country was hard up'.[77] After detailing how during 1963 and the 1964 financing had reignited disagreements between the NIO and civil servants at the Admiralty, which he admitted had upset friends and enemies, Deacon revealed a line of reasoning he had begun to deploy to justify greater investment in the NIO. In a rambling polemical passage of his letter to Cook he argued,

> We are doing magnificent work, which is generally recognised, even in our country, but especially abroad, and we need money to get on.... We send representatives to interminable meetings about the gear that is to go in the Hydrographer's icebreaker oceanographic ship while we cannot put a much needed thing in our own. We give lots of advice about the Hydrographer's three new survey-oceanographic ships and cannot afford to run one which is doing first class work already .... I am not bothering you just out of sheer exasperation...you know the people and departments that are concerned and your advice would be a great help.... It cannot be that the country cannot afford it; our contracts with contractors, the icebreaker-oceanographic ship and the Hydrographer's Oceanographic Branch show that.[78]

These arguments and discussions continued throughout the period and ran in the background of the debates and changes caused by the Trend Report. Deacon's focus on the day-to-day running of the institute distracted him from these upper-level changes. Whereas in 1960 he had been able to contribute to policy discussions, Deacon was now too far from these discussions to be particularly effective. As a "node" Deacon could no longer keep strands of the network together. Taking in the biologists severed the link with the Admiralty, but leaning too much on these connections affected those with the Treasury. It was not because of Deacon's lack of strategic leadership that his network was crumbling, but rather because the evolution of its components meant that he was no longer the central mediating actor between all nodes of his network. Deacon failed to find consensus and was unable to dissuade fellow oceanographers from projecting their own visions for oceanography under the new proposals for the restructuring of science.

The Universe's First Minister for Science[79]

The rise of science policy to a more central position within political debate and Deacon's marginalisation created instability. As argued previously, this took place behind the scenes in a period when oceanography seemed to be happily pursuing national and international collaborative projects. Conversely, back in Britain, four years of civil service re-conceptualisation of the place of science in government was finally coming to fruition and being brought to Parliament. The pivotal year for the shift in science policy in Britain was 1963. The government became pro-active about science policy as the report of Burke Trend's committee was finally returned to Parliament, briefly shining a spotlight onto the work of the Office of the Minister for Science.[80] In response to this, the Royal Society, university scientists under the Department for Scientific and Industrial Research (DSIR), and those such as Deacon in the defence-led research community felt especially threatened. Meanwhile their rivals saw this as an opportunity to regain some "clout" from their adversaries who had the upper hand under the civil science system.

Just as there was change in the air for science policy in government, so too there had been significant political changes. The Labour party emerged rejuvenated by a new leader, Harold Wilson, who propagandised a vision of a new British age, forged in the 'white heat' of scientific revolution.[81] This created a climate in which both the Conservatives and Labour party were keen to show their scientific modernising credentials to the electorate while—essentially—seeking to cut spending for defence science.[82] This made the implementation of the recommendations of the Trend Report an important imperative for government. Science became a primary policy debate in the latter half of 1963. This was the catalyst for rapid changes in science policy and administration. However, debates existed at various levels from Cabinet through the civil service, national scientific societies and committees, scientific institutions, and between individual scientists. And even though the Royal Society attempted to present something that looked like scientific consensus, the Minister for Science, Lord Hailsham, appeared to have no enthusiasm for taking their views seriously.

Lord Hailsham (Quintin Hogg), Minister for Science from 1959 until 1964, began to shape a science policy of his own. The ideas Hailsham had for science policy in 1960 were later published in *Science and Politics*. In this book Hailsham stated: 'we must realise at the outset that in this battle of ideas in which we are engaged, science and technology are benevolent neutrals, willing to ally themselves with either side which seeks their aid.'[83] The issue was in the details of just how one might make science and technology ally to Britain's benefit. Nevertheless, Hailsham offered a voice and sympathetic ear to certain members of the British scientific community and university scientists in particular. This prompted the establishment of the Trend Committee to investigate government-funded civil science and future policy towards its governance.[84] This stemmed from the notion that the DSIR could no longer cope with the volume of government-supported science. As science was increasingly being seen as a social resource that contributed to Britain's economic prosperity, the efficient management of science was a major government priority.[85] This report was commissioned alongside the report of the Committee on Higher Education chaired by Lord Robbins, and known as the Robbins report, which was tasked with investigating higher education; both reports were published in Parliament in 1963.

As Britain's first Minister for Science, Lord Hailsham had been concerned about the lack of co-ordination and symmetry, and the overall place of science within the British state. He saw that science in the UK was conducted either in universities or at governmental institutions, many of which were a legacy of the Second World War. These government research establishments were collectively termed civil science. Soon after coming to office Hailsham published his philosophy on science and the state:

> I would claim that if Government sought to interfere with the scientist in the absolute freedom of his explorations, and the integrity and independence of his speculation, it would ultimately destroy his real source of vigour, or alternatively frustrate his purpose by undermining his confidence and his will to co-operate.[86]

When Hailsham became Minister for Science most government funding for science came out of the defence budget; he claimed in 1960 that three-quarters of spending on science was provided from defence funds. This implied that most civil science establishments were funded for defence purposes, including the NIO. Hailsham was particularly wary of this relationship. He warned: 'the marriage between science and defence is corrupting, and will at best turn science from a liberating to a destructive force, and at worst ultimately dry up the wells of inventiveness in the scientist himself.'[87] It should also be noted that civil science that had demonstrated little actual benefit to immediate defence concerns, or could be just as easily conducted within university environments, offered the opportunity to slim down central government, a primary Conservative objective. Trend was tasked with finding evidence for Hailsham's ideological position and suggesting ways in which science could be moved away from the defence budget and placed in the care of the universities.

Trend concluded that the system employed by the British government to manage research in national institutes was flawed, outdated, and unfit for purpose. The report paved the way for the establishment of national research councils for the individual sciences. These new research councils came into existence during 1965, following the 1964 general election.[88] Although it is undeniable that 'at the 1964 General Election, science became a main platform and a rallying cry for both political parties', this unintentionally created cross-party support for an overhaul of British science policy.[89] Prompted by the commissioning of a report, other non-government actors began to propose and formulate their own reports on the future for ocean sciences in Britain. The growing concern about budgetary issues forced Deacon to take decisions that, far from restoring links in the network, alienated other prominent "nodes" of this web of relations. Deacon tried to prevent the network from collapsing as he coped with external pressures; however in doing so he involuntarily contributed to its disintegration.

George Deacon and the NIO were not the only UK actors undertaking oceanographic research within military circles. Edward Bullard and his assistants at the Department of Geodesy and Geophysics at the University of Cambridge had been conducting research aboard Royal Navy vessels since the 1930s. Bullard's work, however, was not directly funded by the Admiralty, but operated within the framework of funding for university science, managed by the DSIR and the University Grants Committee. Deacon's and Bullard's careers had developed in parallel; by 1960 both were well known both within the UK and internationally. Similarly, their personal scientific standing had risen and both had respectively intertwined themselves inside the Admiralty and the DSIR. Whilst Deacon was advocating expansion of the NIO's programme and negotiating increased funding, Bullard was making separate proposals to the DSIR to raise the profile of university-sponsored oceanography, attempting to wrestle some control of the financing away from the NIO.

In 1960, when Deacon informed Bullard that he would be required to contribute to *Discovery II*'s running costs, Bullard had no qualms about using contacts in Whitehall to apply pressure on Deacon.[90] Bullard's reply to Deacon suggests that Bullard was well informed about the Treasury rules regarding the funding of projects from various sources within the government. He knew that if the Admiralty was funding the vessel then Cambridge could not use DSIR or University Grants Committee money to contribute towards the running of the ship, as the Treasury operated a policy of one project, one funding department or ministry. Bullard now wrote to the Office of the Minister for Science (OMS). The senior secretary dealing with oceanography suggested 'as to finance for Oceanography, I should very much like to have an off-the-record talk sometime soon.... I got the impression that some sort of streamlining may be necessary.'[91] Prior to the establishment of a Ministry for Science, Deacon's opinion had swayed over most government policy concerning oceanography. Bullard's standing as an authority on the opinions of the oceanographic community within Britain began as a counterpoint to Deacon's views. Bullard's resentment was not just a result of the unjustified request for funds for the use of the NIO vessel. At the 1960 conference convened by the BNCOR, the affairs of the universities—in which Bullard had a direct interest—had been brushed aside with the statement that their troubles required separate Royal Society representation to support their own proposals.[92] This did not sit well with Bullard or other university oceanographers who felt Deacon was neglecting or misrepresenting their opinions and views. Therefore when the DSIR decided that a review of oceanographic studies within universities was required in December 1962, Bullard—rather than Deacon—was chosen as the chair for the panel in charge of writing the subsequent report.

The remit given to the DSIR Oceanography Panel was clearly a challenge to Deacon's own programme. The panel was directed to 'give consideration to the ownership and use of research ships since, in view of the present weakness in oceanographic research in this country, it might be desirable for research organisations to own the ships rather than having to collaborate in the programme of the Royal Navy'.[93] Additionally the panel was instructed to keep in 'close touch' with the NATO sub-committee on oceanographic research, UNESCO's Intergovernmental Oceanographic Commission, and SCOR committees, to which Deacon had been the dominant British contributor since their inception. Finally the panel had to include both physical and biological oceanographers. Deacon and the Hydrographer, Vice Admiral Irving, now became the sole non-university members of the Panel, a clear sign of their diminished influence in oceanographic affairs in Britain.

By the time the final report was drafted, Deacon's thoughts had been condensed into a small section. It is unlikely that Deacon saw the final document as anything other than a threat to his perceived custodianship of British oceanography. But the report clearly suggests a "rebellion" by university scientists against the control and dominance of Deacon in particular, and by extension of the NIO. The report itself was long and detailed. Whilst extensive in its praise for Cambridge's Department of Geodesy and Geophysics, the conclusion of the report was clear: 'insufficient facilities, particularly sea going facilities are available for university research workers.'[94] Bullard's report was bold, proposing an expenditure of £4.5 million over five years to be divided into three portions, with £1 million being used to build a deep-sea vessel (when only £360,000 had been spent on the NIO's new ship), £350,000 for shallow-water vessels, and the remainder to cover the associated running costs. This was all conceived as being additional to the NIO, rather than an attempt to either subsume the NIO into the DSIR or to take money away from the NIO. However, whilst the report may have had an impact in 1962 when it was commissioned, by 1964 attitudes towards science policy in Whitehall were heavily influenced by the conclusions of the Trend Report.

The Trend Report had significant implications for oceanography as a discipline. The report had recommended the creation of various research councils, but initially there would be just two: the Science Research Council (SRC) and the Natural Resources Research Council (NRRC), with the earth sciences and oceanography placed under the NRRC. The response of senior oceanographers such as Bullard and Deacon was swift. Their science—they agreed—was not a second-tier discipline and they felt strongly that if oceanography were not included in the SRC then it would not be taken seriously as a discipline. In effect oceanography was now considered in government policy as an applied rather than a theoretical discipline. There were also deeper concerns about classification. In a report to the Office of the Minister for Science, Bullard attempted to construct a list of what disciplines comprised the earth sciences. Bullard was annoyed with his own efforts and wrote, 'This list is a list, it is not a complete logical classification and it is not exhaustive; knowledge, in spite of biblical authority to the contrary, is not a tree, it is a net or a web or perhaps a tangle, everything is connected to everything else and no subject has sharp boundaries.'[95]

The Trend Report suggested placing various government research establishments outside government ministries and handing over control to the new research councils dismantling most government civil science. The implication of this transformation for the NIO seemed to be clear. Bullard argued to the Office of the Minster for Science that the NIO scientific programme should be placed under the control of the research council; however, the research should continue to be funded by the Admiralty.[96]

Meanwhile the new President of the Royal Society, Sir Howard Florey, was concerned about the implications of the Trend Report for British science. His concerns lay in the rejection by the Trend Committee of many of the proposals presented by the Royal Society, primarily the recommendation that a Civil Science Board should be established to overlook the establishment of research councils, the members of this board being independent scientists who would devote half of their time to working for the new board. Creating new boards did not fit the Trend Report's objective of streamlining science. Florey and Sir Maurice Dean—who at the time was the Joint Permanent Secretary of State at the Department of Education and Science—met to discuss the Trend Committee's conclusions early in 1964. At this meeting Dean informed Florey that Hailsham had his own vision for the new research councils. Florey recorded that

> It appeared that the Minister thought that there was a 'critical' size for a research council and that any such council should have spending power in the range of £5–10 million a year; the minister therefore had in mind to extend the field proposed to be covered by the Natural Resources Research Council (NRRC)... and to change the proposed name...to something else, for instance Environmental Science Research Council or Earth and Marine Sciences Research Council.[97]

Within the higher echelons of the Admiralty administration these conclusions were seen as a nuisance rather than a useful simplification of existing practice. The proposals gave the Admiralty's Head of the General Finance Office concerns as early as December 1963, when he wrote a minute conceding that the issues surrounding the NIO and the Trend Report were complex. Since 1949 the Admiralty had handled the affairs of the NIO using a policy of support with limited commitment, that is, a policy of "limited control". However, the Admiralty felt that the proposals of the Trend Report were open to debate, and therefore it was likely that present arrangements, whereby the Admiralty had administrative control, were likely to continue. This control would be limited to representation on the NOC, secondment of staff from the CRNSS, and the granting of funds. On these funds, he stated, there was no day-to-day control 'in the normally accepted sense'.[98] In 1963, the main concern with the NIO remained, namely that the activities of the Institute were sharply divided between the physical and biological, and that maintaining a proper balance between these strands of research was a primary concern. The head of finance explicitly stated that 'while a certain amount of knowledge for defence can be derived from biological research the benefit here is only marginal and our main interest lies on the side of basic physical research'.[99]

In early January 1964 the Permanent Secretaries of various government departments came together to discuss the implications of the Trend Report. This meeting demonstrated that there was a gulf between the opinions of the OMS, scientists, and civil servants. Sir Richard Way, who was one of the youngest Permanent Secretaries, was particularly damning about the report—he described it as 'woolly and harmful to the national interest'.[100] The general opinion of the meeting was that, from the soundings taken by the Permanent Secretaries, most scientific opinion disagreed with Hailsham.[101]

Now the Admiralty changed tack, suggesting that the time for limited control was over. Oceanographic research of military use would be conducted in-house and the NIO left to its own (or the Treasury) devices: 'we believe it would be undesirable to perpetuate divided control of the NIO and that the national interest, both civil and defence, would be best served by putting the Institute under the control of the SRC, removed from Navy Votes and financed by the OMS.'[102] The OMS opened a file entitled 'Discussion on who should have responsibility for oceanography' which looked at the issues surrounding the removal of the NIO budget from the Admiralty. In accepting that the NIO would be best placed under a research council, the Admiralty appeared to be supporting Hailsham's pet project; however it should be noted that this solution seemingly allowed the Admiralty to rid itself of a financial burden whose value had come to be increasingly questioned, partly because of the diminished role of oceanographic studies in the creation of the British-made sea surveillance system.[103]

During subsequent discussions between the Royal Society and the newly established Department of Education and Science (DfES), it emerged that there were no major obstacles to the NIO joining the Natural Environment Research Council, the name of which had also been revised at a meeting on 23 April 1964 to accommodate research beyond natural resources. Over the summer little changed, however, due to a long-standing civil service tradition that the files of the previous government should not be made available to the new government. When the Labour party returned to power following the general election in October 1964, they were not to see where agreement lay. Instead a new file was started by the civil servants in DfES, entitled 'NERC Future of National Oceanographic Council & National Institute of Oceanography.' This subtle change allowed for a shift in emphasis: now the NERC, the NOC, and the NIO were inextricably linked. This change had major implications for Deacon who was not to play a leading role in the new set-up.

Harold Wilson's Labour government, which swept to power in October 1964, had been elected on a manifesto pledge to reinvigorate British science and industry, so as to bring about the promised economic growth and prosperity which had been a central policy ambition of all political parties since the war. With Labour coming to office, there seemed an inevitability that new research councils were going to come into existence. Lord Alexander Robertus Todd, as chair of the Advisory Committee on Science Policy (ACSP), had been an important conduit for scientists to have their opinions heard within Whitehall. As a long-standing chairman of the ACSP (1952–64) he was a well-established figure in science policy circles, and commanded a degree of respect as a scientific administrator. After a meeting with Bullard, Todd relayed the content of the discussion straight to senior civil servants in the DfES. Bullard had informed Todd that he held the 'strongest possible view' against Deacon being made chairman of the newly formed NERC Oceanographic Committee.[104]

Bullard need not have feared, as the DfES intended to appoint Sir Graham Sutton instead of Deacon. Sutton was not an oceanographer, but had been director general of the Meteorological Office since 1953, a position comparable with the NIO directorship. However Sutton, unlike Deacon, was seen as somebody who would not 'set the Thames on fire'.[105]

This was a strong statement considering Deacon's long-standing relationship with government. However, his provocative—yet compromising—attitude, which had worked well during the immediate post-war years, was now a hindrance rather than an asset. His position also appeared much weaker than before; with arrangements for the NIO undecided, this was the opportunity for others who harboured their own ambitions about the direction of ocean science policy to make a move against Deacon's hegemony. These tensions at home affected Deacon's reputation internationally.

By 1964 leading US oceanographer Roger Revelle suggested that he no longer trusted the NIO director to lead the British contingent. Revelle wrote to Bullard: 'we must get British oceanography out of Deacon's hands and into those of someone with a broader and more liberal view, and particularly must associate it more closely with the universities.'[106] Hamblin suggests that Bullard demurred on this point, and felt a policy change in the air.[107] Bullard genuinely believed that the DSIR report would force through change, but his closeness to its authorship meant that he failed to foresee that it might fail to be accepted wholeheartedly by the incoming NERC committee. Bullard soon moved against Deacon, informing Todd that Deacon would be a poor choice as NERC chairperson.[108]

Deacon was also losing ground with other important components of his network. To those in the Admiralty, Deacon had become increasingly difficult to deal with. Many thought he had proven irrational, often too emotional, and certainly over-opinionated and dominant.[109] For years officials had taken to having pre-meetings to agree the line that they were going to take with Deacon, so that they could be firm and prepared for his arguments, some of which they privately conceded were valid but nonetheless beyond what they were able to accept.[110] The Admiralty managers had turned a blind eye to Deacon's idiosyncrasies because of the importance of the oceanography programme for their own surveillance and operational tasks. But its reduced significance now made it possible for them to be more direct. When Deacon wrote one too many venomous letters, an Office of the Minister for Science official wrote, 'it is time that Dr. Deacon discovered where and how most appropriately he should formulate and state his case.'[111]

During these high-level deliberations, in which Deacon had been increasingly sidelined, he had failed to keep the staff at the NIO informed about developments. Concerned with what was happening to their institute, senior NIO scientists wrote to Sutton. In their letter they voiced their concerns: 'most of our disquiet almost certainly arises from ignorance. We hear various rumours about NERC – who is to be on it, how it will work, and so on.'[112] Sutton replied to the senior staff that as the NERC was not yet constituted he did not want to be drawn into committing a definite statement on how its oceanography committee would be composed.[113] However he did write personally to Deacon to attempt to smooth any ruffled feathers that his appointment may have caused.[114]

By January 1965 the pathways towards the formation of the NERC were beginning to take shape. The new government had decided to combine their own manifesto pledges for science and technology with the Conservative schemes for research councils and reorganisation of the governance of science by Whitehall. The new bill, the Science and Technology Act (1965), passed through Parliament on 23 March 1965, following months of careful planning. In the end Deacon was not completely demoted, as it had become apparent in January that the new bill would not dissolve the National Oceanographic Council.[115] The civil service statement to the NOC informed oceanographers that continuing the NOC would 'run counter to the whole concept of the Government's re-organisation'.[116] Eventually two long-standing members of the NOC submitted papers to a DfES civil servant, who however dismissed these as 'falacious (sic)' and 'irrelevant' respectively, thus showing disappointment that the NOC continued.[117]

Throughout the spring of 1965 the passing of the Science and Technology Act consumed the energies of the DfES civil servants. However, after 1 June when the provisions of the Act came into force, placing staff at the NIO in the employment of the NERC rather than the Royal Navy, the ministry began to deal with the remaining anomalies. Handling the winding-up of the NOC was a particularly delicate operation. The research and development section of the MoD wrote to the DfES, suggesting that this matter needed to be 'handled carefully'. The MoD was so concerned about the reaction of the single-minded oceanographers on the NOC that they emphasised that 'some very careful groundwork is necessary…this is being handled personally by the Under Secretary of State (Royal Navy)'.[118] That civil servants were worried about how smooth this transition could be showed the limits of their power to compel oceanographers to follow direct instructions, and the continued need for consensus, even if the civil servants were aware that the scientists whose careers they were affecting seldom embraced change. Surprisingly, the

NOC accepted the change and suggested that there should be some publicity to mark the 'demise of the Council'.[119]

This was not as straightforward as it may have seemed and the DfES was fully aware that press attention had to be carefully managed. They noted the MoD suggestion that the occasion should be positive and promote the NERC and the DfES to show how the government was looking to the future role of oceanography. As the months passed and the procedural motions moved slowly forward to official acceptance of the petition, a suggestion was made that the presentation should be made to the press at the meeting of the final NOC to be held on 20 October at the NIO. Latent problems immediately arose. The plan for the day was to formally wind up the NOC in a meeting in the morning before holding a presentation, which was described as a novel 'Americanism' supposedly 'beloved in the Ministry of Defence'.[120] This was a risky proposal, as not all oceanographers agreed with the NOC disbandment. As the MoD pointed out, 'There is, of course, an obvious risk involved in courting publicity in this way – e.g. if the Press came across disgruntled members of the NOC or staff of the NIO uncertain about their future – but in my judgement the risk is acceptable.'[121] In an attempt to mitigate this risk, DfES issued a two-paragraph press release mentioning only the future of the NIO under the NERC. In the event it would seem from the *Statist* article quoted at the opening of this chapter, and from other similarly negative press, that the dissenting opinions—which the MoD feared would come from the oceanographers themselves—had indeed leaked out.

Change in Whitehall

This chapter has charted the movement of British oceanography from military to civilian patronage in the early 1960s. The process of accomplishing this was infinitely more complex than the establishment of the NIO had been under the Attlee government during the 1940s. This was as a result of the grossly enlarged government machine, and the increasing centralisation of government science policy. It was also as a result of the changes to NATO's military strategy and policies, in addition to a general shift in the perception of what was suitable research for civilian scientists to be undertaking and what was suitable only for military scientists. This arbitrary boundary between civilian and military was seen by Deacon as needless because there was so much scientific overlap between the two "strands" of oceanography, but to the military the demarcation was fundamental.

When the collapse of confidence in Deacon in Whitehall came, it happened swiftly. It can be said to have begun in 1963 and was merely confirmed by the new Labour government in 1965. Before 1963, the NIO had been growing steadily, building research programmes and new survey vessels. After 1965 it represented just one of many research establishments in the country bidding for funding from a centralised fund.

The experience of oceanographic sciences in Britain during the so-called "technocratic moment" has a very different trajectory seen through its public image rather than in the reality of the corridors of power. To many within these corridors, oceanography appeared as a key component of Cold War defence science, as shown in Chap. 5; however, as shown here the funding to undertake civilian-military projects and expand the Institute was, to Deacon, a political battlefront rather than something to be taken for granted. Whilst some sciences certainly can be said to have flourished under a technocratic modernism visible in aviation, civil engineering, and perhaps space technologies, oceanographers struggled as their fortunes were closely tied to those of their patrons. And whilst the Royal Navy might have finally won the nuclear initiative from the RAF with Polaris, this was a small concession, which did not benefit the Royal Navy as a whole.[122]

The Trend Report and the subsequent restructuring of civil science have been interpreted in many ways. Arguments as to why civil science and defence science needed to be separated ranged from the corrupting influence of militarising science to attempts to differentiate the defence budget to make it appear to be reducing in size and, finally, to an attempt to place science back into the hands of the scientists. Deacon feared this the most. He understood that scientists each individually believed themselves to be experts in their own right, and a voice for their own specific disciplines. On the other hand, it was comparatively easy to sell the idea that science could herald advantages to military officials dealing with an increasingly technical battlefield, where inter-service rivalry was ever present. Ultimately the debates of the early 1960s came down to questions of what constituted the right scientific programmes and who was the right body to fund such a programme.

Whether or not Deacon actively sought out a physical programme to the detriment of a biological programme is difficult to determine. Mackintosh's "defection" to the Whale Research Unit at the Natural History Museum was opportune and ultimately the biological programme did not expand at the same rate as the physical programme at this time. Additionally, Deacon's relationship with Bullard and the Admiralty became increasingly strained as he had to fight to defend his institution; ultimately this leached into the international sphere where Deacon was no longer the only British oceanographer with a reputation for getting good science done—as shown by Revelle's remarks. He could no longer be the man sitting at the centre of a network representing military oceanography, as the science itself was becoming more publicly engaged through the work of people like Jacques Cousteau, and at the same time the security dilemma of ballistic missile submarines pushed military scientific programmes under a veil of increasing secrecy. One unanticipated consequence of the events of 1965 was that Deacon decided that the NIO needed him to remain at its helm to steer it through the coming changes. Even though Deacon should have retired that year he was to stay with the NIO until 1971. As the next chapter will show, the last five years of Deacon's "reign" at the NIO were typified by an attempt to diversify and "civilianise" the research programme so as to please new patrons. The results of these attempts were, unsurprisingly, mixed.

Notes

1. 'The sea is whose oyster?', *The Statist* (29 October 1965).
2. 'The Sea is Whose Oyster?', *The Statist* (29 October 1965). The article claimed 'half a million pounds a year is all Britain spends on oceanography; Canada spends £25m., the United States £100m. and the Russians probable even more'.
3. 'The Sea is Whose Oyster?', *The Statist* (29 October 1965).
4. Frederick Brundrett, "The Neglected Sea," (The Twenty-Ninth Haldane Memorial Lecture, delivered at Birkbeck College, London, 7 March 1963). Churchill College Archives Centre, Sir Frederick Brundrett papers, Churchill/BRUN 2/1.
5. David Edgerton, "C.P. Snow as Anti-historian of British Science: Revisiting the Technocratic Moment, 1959–1964," *History of Science* (2005): 187.
6. On C.P. Snow see Guy Ortolano, *The Two Cultures Controversy: Science, Literature and Cultural Politics in Postwar Britain* (Cambridge: Cambridge University Press, 2011).
7. Samuel H. Beer, *Britain Against Itself: The Political Contradictions of Collectivism* (New York: Norton, 1982).
8. Edgerton, "C.P. Snow as Anti-historian of British Science," 187.

9. On Britain's space programme, see Stuart Butler, "National prestige and in(ter)dependence: British space research policy, 1959–73" (Ph.D diss., University of Manchester, 2017).
10. Jim Tomlinson, "Conservative Modernisation, 1960–64: Too little, too late?," *Contemporary British History,* 11:3 (1997): 18–38; Jon Agar, Brian Balmer, "British Scientists and the Cold War: The Defence Research Policy Committee and Information Networks, 1947–1963," *Historical Studies in the Physical and Biological Sciences,* 28:2 (1998): 246; Jon Agar, "The New Price and Place of University Research: Jodrell Bank, NIRNS and the Context of post-war British Academic Science," *Contemporary British History,* 11:1 (1997): 23–24.
11. Quoted in 'Notes and Comments', *New Scientist* (13 August 1959): 173.
12. *New Scientist,* (13 August 1959). *New Scientist* began in 1956 as a weekly journal of science news.
13. 'Notes and Comments', *New Scientist* (13 August 1959).
14. See Chap. 3 of this book.
15. Letter Brundrett to Deacon, 12 August 1959, GERD Papers, D1/7, NOC Library (Southampton).
16. Letter Brundrett to Deacon, 15 October 1959, GERD Papers, D1/7, NOC Library (Southampton).
17. Letter Deacon to Brundrett, 13 August 1959, GERD Papers, D1/7, NOC Library (Southampton); Letter from Deacon to Swallow, 10 March 1960, GERD Papers, D4/4, NOC Library (Southampton).
18. Draft article by Frederick Brundrett attached to letter sent to George Deacon, 15 October 1959, GERD Papers, D1/7, NOC Library (Southampton).
19. Helen M. Rozwadowski, "Arthur C. Clarke and the Limitations of the Ocean as a Frontier," *Environmental History* (2012): 588.
20. Arthur C. Clarke, *The Challenge of the Sea* (New York: Holt, Rinehart and Winston, 1960); Arthur C. Clarke, *A Fall of Moondust* (London: Gollancz, 1961).
21. Gary Kroll, *America's Ocean Wilderness: A Cultural History of Twentieth-Century Exploration,* (Kansas: University of Kansas Press, 2008): 171–8.
22. Bradford Matsen, *Jacques Cousteau: The Sea King* (London: Vintage Books, 2010).
23. Rozwadowski, "Arthur C. Clarke," 589.
24. Kroll, *America's Ocean Wilderness,* 174–5.
25. For parallels between US and European science in the Cold War see John Krige, *American Hegemony and the Postwar Reconstruction of Science in Europe* (Cambridge, MA: MIT Press, 2006).
26. Arthur C. Clarke, *The Deep Range* (London: Frederick Muller, 1957).

27. 'Marine Science – Why Not more of it!' draft article by F. Brundrett, undated, GERD Papers, D1/7, NOC Library (Southampton).
28. Ibid.
29. Letter from Brundrett to Deacon, 21 October 1959, GERD Papers, D4/4, NOC Library (Southampton).
30. Deacon's introduction was entitled, "The increasing significance of marine science," *New Scientist* (13 August 1959): 178–9; see also "Notes and Comments," *New Scientist*, (13 August 1959): 173.
31. For more on the Royal Society in the Cold War, see Jennifer R. Goodare, "Representing Science in a Divided World: The Royal Society and Cold War Britain," (Ph.D diss., University of Manchester, 2013).
32. Royal Society Minutes of Council (21 May 1959): 238–9; Royal Society Minutes of Council (18 June 1959): 254.
33. See Chap. 3.
34. Letter from D.C. Martin to Deacon, 14 January 1960, GERD Papers, G1/1, NOC Library (Southampton).
35. Goodare, "Representing Science in a Divided World," 120.
36. Jacob D. Hamblin, *Oceanographers and the Cold War: Disciples of Marine Science* (Seattle: University of Washington, 2005): 101–104.
37. Jeff Hughes, "Doing Diaries: David Martin, the Royal Society and scientific London, 1947–1950," *Notes Rec. R. Soc.*, 66:3 (2012): 273–294.
38. Goodare, "Representing Science in a Divided World," 120.
39. Letter from D.C. Martin to Deacon, 14 January 1960, GERD Papers, G1/1, NOC Library (Southampton): 'The Physical Secretary and I would very much like to have a talk with you before the meeting of the British National Committee for Oceanic Research…perhaps you could come then and take lunch with us before going on to the meeting of the National Committee.'
40. Letter from D.C. Martin to Deacon, 23 January 1960, GERD Papers, G1/1, NOC Library (Southampton).
41. Goodare, "Representing Science in a Divided World," 120–123, 144–156.
42. Harold Macmillan, "Tercentenary Banquet, Grosvenor House, Tuesday 26th July 1960: The Toast of the Royal Society Proposed by the Rt. Hon. Harold Macmillan, M.P.," *Notes and Records of the Royal Society*, (1961): 31–37.
43. Goodare, "Representing Science in a Divided World," 145.
44. Peter Collins, "A role in Running UK Science?," *Notes and Records of the Royal Society* (2010): 122.
45. Minutes of the meeting of the Council of the Royal Society held on 5 May 1960, 3 November 1960, 15 December 1960; 404, 483, 502–3, contained within Royal Society Minutes of Council 1957–61, Vol. 20.

46. Letter from D.C. Martin to G.E.R. Deacon, 6 September 1960, GERD Papers, G1/12, NOC Library (Southampton).
47. BNCOR, 'Need for expansion in marine science by the United Kingdom', Report of Conference held on 11 November 1960, CAB 124/2930, TNA (London).
48. Ibid.
49. The Meeting was held on 11 November 1960.
50. Letter from Deacon to Willis, 17 July 1961, GERD Papers, D1/7, NOC Library (Southampton).
51. Letter from Deacon to Willis, 19 July 1961, GERD Papers, D1/7, NOC Library (Southampton).
52. Letter from Deacon to Willis, 17 July 1961, GERD Papers, D1/7, NOC Library (Southampton).
53. Robert Self, *British Foreign & Defence Policy since 1945: Challenges & Dilemmas in a changing world* (London: Palgrave Macmillan, 2010): 162.
54. Sir Gerald Templer (Chief of the General Staff) was said to have informed Sandys, 'Duncan, you're so bloody crooked, that if you swallowed a nail, you'd shit a corkscrew!; Dennis Healey, *The Time of My Life* (London: Michael Joseph, 1989): 257.
55. Ritchie Ovendale, "Macmillan and the Wind of Change in Africa, 1957–1960," *The Historical Journal* 38:3 (1995): 455–477.
56. Defence White Paper, 1960 (8 February, 1960) Cmnd. 952., CAB 129/100/14, TNA (London).
57. Orr-Ewing had been in post less than a year (previously holding the office of the Parliamentary and Financial Secretary to the Admiralty between 1957 and 1959).
58. Note attached to Report on the NIO by J.S.L. for the Civil Lord of the Admiralty, 17 August 1960, ADM 1/27490, TNA (London).
59. Report on the NIO, for the Civil Lord of the Admiralty, 17 August 1960, ADM 1/27490, TNA (London).
60. Note attached to Report on the NIO by J.S.L. for the Civil Lord of the Admiralty, 17 August 1960, ADM 1/27490, TNA (London).
61. 'The Sea is Whose Oyster?' *The Statist* (29 October 1965).
62. Letter from Deacon to Brundrett, 19 October 1959, GERD Papers, D1/9, NOC Library (Southampton).
63. Letter from W.N. Smith (H.M. Treasury) to A.E. Martin (Admiralty, Civil Establishments Branch), 8th July 1960, T 213/820, TNA (London).
64. Letter from Martin to Smith, 29 August 1960, T 213/820, TNA (London).

65. Letter from Deacon to Lythall (Chief Scientist, Admiralty Underwater Weapons Establishment), 27 June 1961, GERD Papers, D4/4, NOC Library (Southampton).
66. 'Inspection of the National Institute of Oceanography – October 1960, A note on the work of the Institute as seen by a member of the Inspection Team', T 213/820, TNA (London).
67. Anthony Laughton, interviewed by Paul Marshal, *British Oral History of Science Interview,* C1379/29 Track 5, p.106 of Transcript, http://sounds.bl.uk/relatedcontent/TRANSCRIPTS/021TC1379X0029XX-0000A0.pdf.
68. 'Report on the Treasury/Admiralty inspection of the National Institute of Oceanography', October, 1960, T 213/820, TNA (London).
69. Letter from R.G. Williams to R. Devereux, 1 December 1960, T 213/820, TNA (London).
70. Graham D. Burnett, *The Sounding of the Whale: Science and Cetaceans in the Twentieth Century* (Chicago: University of Chicago Press, 2012): 387.
71. As seen in Chap. 2, Mackintosh as Director of the Discovery Investigations had overseen the early career of Deacon, before the Admiralty promoted Deacon to Director of the NIO putting Mackintosh in a subordinate position as Assistant Director.
72. Burnett, *The Sounding of the Whale,* 465–6.
73. Memo by Mackintosh sent to Jourdain, 20 Feburary 1948 received in Cabinet office 1 March 1948, CAB 124/555, TNA (London).
74. Letter from Deacon to Willis, 24 January 1961, GERD Papers, D4/4, NOC Library (Southampton).
75. Self, *British Foreign & Defence Policy since 1945,* 165.
76. Report on the NIO to the Civil Lord of the Admiralty, 17 August 1960, ADM 1/27490, TNA (London).
77. Letter from Deacon to Sir William Cook, 21 July 1964, GERD Papers, D4/6, NOC Library (Southampton).
78. Letter from Deacon to Sir William Cook, 21 July 1964, GERD Papers, D4/6, NOC Library (Southampton).
79. The Universe's first minister was Hailsham's own self-styling.
80. Tomlinson, "Conservative modernisation," 28.
81. David Edgerton, "The 'White Heat' Revisited: The British Government and Technology in the 1960s," *Twentieth Century British History,* 7:1 (1996): 53–82; Richard Coopey, "Industrial Policy in the White Heat of the Scientific Revolution", in *The Wilson Years,* ed. Richard Coopey, Steven Fielding, and Nick Tiratsoo (London: Pinter, 1992).
82. M. W. Kirby, "Blackett in the "White Heat" of the Scientific Revolution: Industrial Modernisation under the Labour Governments, 1964–1970,"

The Journal of the Operational Research Society 50:10 (1999): 985–993; Richard Findley, "The Conservative Party and Defeat: the Significance of Resale Price Maintenance for the General Election of 1964," *Twentieth Century British History* 12:3 (2001): 327–353.

83. Viscount Hailsham, *Science and Politics* (London: Faber & Faber, 1963): 78.
84. David Edgerton, "The 'White Heat' Revisited: The British Government and Technology in the 1960s," *Twentieth Century British History* 7:1 (1996): 53–82.
85. Sabine Clarke, "'A Technocratic Imperial State?' The Colonial Office and Scientific Research, 1940–1960," *Twentieth Century British History* 18:4 (2007): 453–480; Guy Ortolano, *The Two Cultures Controversy: Science, Literature and Cultural Politics in Postwar Britain* (Cambridge: Cambridge University Press, 2011).
86. Hailsham, *Science and Politics,* 14.
87. Hailsham, *Science and Politics,* 15.
88. Other research councils had already been established in the UK prior to the post-Trend expansion of the research council system: the DSIR (1916), Medical Research Council (1920) and the Agricultural Research Council (1931).
89. Tom Wilkie, *British Science Policy Since 1945* (Oxford: Oxford University Press, 1991): 19.
90. Letter from Bullard to Miss Morris (DSIR), 28 September 1960, CAB 124/2930, TNA (London).
91. Letter from Quirk to Bullard, 30 September 1960, CAB 124/2930, TNA (London).
92. Report of Conference, 'British National Committee for Oceanic Research Need for Expansion in Marine Science by the United Kingdom', 11 November 1960, CAB 124/2930, TNA (London).
93. DSIR Report of the Oceanography Panel: on the Provision of Vessels and other Oceanographic Facilities, 31 January 1964, CAB 124/3037, TNA (London).
94. Note on the Report of the Oceanography Panel Submitted with Sir Harry Melvile's Minute, 25 June 1964, CAB124/3037, TNA (London).
95. 'The Earth Sciences and the Trend Report', 7 February 1964, ECB Papers, E159, E (Cambridge).
96. Ibid.
97. Trend/1 (64), minutes of meeting held 2 April 1964, ECB Papers, E159, CAC (Cambridge).
98. Note of the Head of G.F.O., 4 December 1963, ADM 1/28555, TNA (London).
99. Ibid.

100. Minutes of Permanent Secretaries Meeting, 9 January 1964, ADM 1/28555, TNA (London).
101. Minutes of Permanent Secretaries Meeting, 9 January 1964, ADM 1/28555, TNA (London).
102. Memo for the Deputy Secretary, 17 December 1963. ADM 1/28555, TNA (London).
103. See Chap. 5.
104. Unsigned Minute, 15 January 1965, CAB 124/1823, TNA (London).
105. Ibid.
106. Letter from Revelle to Bullard, 25 February 1964, ECB Papers, C.18, CAC (Cambridge).
107. Hamblin, *Oceanographers and the Cold War,* 186.
108. Minute by Miss Senior, 15 January 1965, CAB 124/1823, TNA (London). Lord Todd won the Nobel Prize for Chemistry in 1957, was chair of the ACSP 1952–64 (served on ACSP 1947–64), and was elected PRS 1975–80; see Brown, D. M., and Kornberg, H., 'Alexander Roberus Todd, O.M., Baron Todd of Trumpington. 2 October 1907–10 January 1997', *Biogr. Mems. Fell. R. Soc.,* 46 (2000) 515–532.
109. Letter from Deacon to Willis, 24 January 1961, ADM 1/27490, TNA (London); Note by Evans 23 October 1964, CAB 124/2178, TNA (London).
110. Note by Evans 23 October 1964, CAB 124/2178, TNA (London).
111. Note by Evans written on a letter from Jones (Admiralty) to Fraser (Treasury), 16 March 1964, (note written 18 March 1964), CAB 124/2178, TNA (London).
112. Letter from Charnock to Sutton, 25 March 1965, CAB 124/1823, TNA (London).
113. Letter from Sutton to Charnock, 29 March 1965, CAB 124/1823, TNA (London).
114. Letter from Sutton to Deacon, 31 March 1965, CAB 124/1823, TNA (London): 'I think you need have no fears regarding the activities of NERC vis-à-vis the NIO. I have taken into account all that you said at our meeting.'
115. Letter from Michael Cary (Ministry of Defence) to Sir Frank Turnbull, (DfES), 20 January, 1965, CAB 124/1823, TNA (London).
116. Draft of Statement to be given to the meeting of the NOC on 29 January 1965, memo 'Organisation for Oceanography', dated 26 January 1965, CAB 124/1823, TNA (London).
117. Letter from Sir Frank Turnbull (DfES) to Sutton, 19 March 1965, CAB 124/1823, TNA (London).
118. Letter from Jones, (Research and Development Finance Division (Naval), MoD) to Hidges, DfES, 8 June 1965, CAB 124/1823, TNA (London).

119. Letter from Hodges to Jones, 23 July 1965, CAB 124/1823, TNA (London).
120. Minute from MoD for the Minister of State for Education and Science (Lord Bowden), 27th September 1965, CAB 124/1823, TNA (London).
121. Minute from MoD for the Minister of State for Education and Science (Lord Bowden), 27 September 1965, CAB 124/1823, TNA (London).
122. On Polaris see Peter Hennessy and James Jinks, *The Silent Deep: The Royal Navy Submarine Service since 1945* (London: Penguin Books, 2015).

References

Agar, Jon. "The New Price and Place of University Research: Jodrell Bank, NIRNS and the Context of Post-war British Academic Science." *Contemporary British History* 11 (1997): 23–24.

Agar, Jon, Brian Balmer. "British Scientists and the Cold War: The Defence Research Policy Committee and Information Networks, 1947–1963." *Historical Studies in the Physical and Biological Sciences* 28 (1998): 209–252.

Beer, Samuel H. *Britain Against Itself: The Political Contradictions of Collectivism.* New York: Norton, 1982.

Burnett, Graham D. *The Sounding of the Whale: Science and Cetaceans in the Twentieth Century.* Chicago: University of Chicago Press, 2012.

Butler, Stuart. "National Prestige and In(ter)dependence: British Space Research Policy, 1959–73." PhD diss., University of Manchester, 2017. British Library EthOS (uk.bl.ethos.713607).

Clarke, Arthur C. *The Deep Range.* London: Frederick Muller, 1957.

Clarke, Arthur C. *The Challenge of the Sea.* New York: Holt, Rinehart and Winston, 1960.

Clarke, Arthur C. *A Fall of Moondust.* London: Gollancz, 1961.

Clarke, Sabine. "'A Technocratic Imperial State?' The Colonial Office and Scientific Research, 1940–1960." *Twentieth Century British History* 18 (2007): 453–480.

Collins, Peter. "A Role in Running UK Science?" *Notes and Records of the Royal Society* (2010): 1–12.

Coopey, Richard. "Industrial Policy in the White Heat of the Scientific Revolution." In *The Wilson Years.* Edited by Richard Coopey, Steven Fielding, Nick Tiratsoo. London: Pinter, 1992.

Deacon, George E.R. "The Increasing Significance of Marine Science." *New Scientist.* 13 August 1959.

Edgerton, David. "The 'White Heat' Revisited: The British Government and Technology in the 1960s." *Twentieth Century British History* 7 (1996): 53–82.

Edgerton, David. "C.P. Snow as Anti-historian of British Science: Revisiting the Technocratic Moment, 1959–1964." *History of Science* 43 (2005): 187–208.
Findley, Richard. "The Conservative Party and Defeat: The Significance of Resale Price Maintenance for the General Election of 1964." *Twentieth Century British History* 12 (2001): 327–353.
Goodare, Jennifer R. "Representing Science in a Divided World: The Royal Society and Cold War Britain." PhD diss., University of Manchester, 2013.
Hailsham, Viscount. *Science and Politics*. London: Faber & Faber, 1963.
Hamblin, Jacob D. *Oceanographers and the Cold War: Disciples of Marine Science*. Seattle: University of Washington, 2005.
Hennessy, Peter, James Jinks. *The Silent Deep: The Royal Navy Submarine Service Since 1945*. London: Penguin Books, 2015.
Hughes, Jeff. "Doing Diaries: David Martin, the Royal Society and Scientific London, 1947–1950." *Notes and Records of the Royal Soceity* 66 (2012): 273–294.
Kirby, M.W. "Blackett in the "White Heat" of the Scientific Revolution: Industrial Modernisation Under the Labour Governments, 1964–1970." *The Journal of the Operational Research Society* 50 (1999): 985–993.
Krige, John. *American Hegemony and the Postwar Reconstruction of Science in Europe*. Cambridge, MA: MIT Press, 2006.
Kroll, Gary. *America's Ocean Wilderness: A Cultural History of Twentieth-Century Exploration*. Kansas: University of Kansas Press, 2008.
Macmillan, Harold. "Tercentenary Banquet, Grosvenor House, Tuesday 26th July 1960: The Toast of the Royal Society Proposed by the Rt. Hon. Harold Macmillan, M.P." *Notes and Records of the Royal Society* (1961): 31–37.
Matsen, Bradford. *Jacques Cousteau: The Sea King*. London: Vintage Books, 2010.
New Scientist. "Notes and Comments." 13 August 1959.
Ortolano, Guy. *The Two Cultures Controversy: Science, Literature and Cultural Politics in Postwar Britain*. Cambridge: Cambridge University Press, 2011.
Ovendale, Ritchie. "Macmillan and the Wind of Change in Africa, 1957–1960." *The Historical Journal* 38 (1995): 455–477.
Rozwadowski, Helen M. "Arthur C. Clarke and the Limitations of the Ocean as a Frontier." *Environmental History* (2012): 578–602.
Self, Robert. *British Foreign & Defence Policy Since 1945: Challenges & Dilemmas in a Changing World*. London: Palgrave Macmillan, 2010.
The Statist. "The Sea Is Whose Oyster?" 29 October 1965.
Tomlinson, Jim. "Conservative Modernisation, 1960–64: Too Little, Too Late?" *Contemporary British History* 11 (1997): 18–38.
Wilkie, Tom. *British Science Policy Since 1945*. Oxford: Oxford University Press, 1991.

CHAPTER 7

New Frontiers of Oceanology and "Environmentalism"

The position and role of science in society changed dramatically in the late 1960s. Robert Cockburn, who had worked on radar countermeasures during the Second World War, suggested in 1967 in an article for the *Journal of the Royal Society of Arts* that often the same area of science could be put to different uses, prioritising either military prowess or peaceful developments. He argued:

> To many people it is a matter of concern that scientific knowledge stemming from a search for truth and understanding should so often find its first application in supporting our less creditable aggressive instincts. But Science can be used either to restrain or promote warfare, to achieve wisdom or to pamper our desires; the choice lies not with the scientists or soldier but with society as a whole.[1]

The emphasis placed on either of these priorities increasingly came to be seen as defining the UK's international political posture.[2] Ocean science, just as in the case of other sciences, had to find its place in a "sea" of politico-scientific affairs.[3] With the re-configuration of British civil science that followed the Trend and Robbins reports (1963) the ties between the National Institute of Oceanography (NIO) and the Admiralty had been weakened. The Institute was now under the control of the newly established Natural Environment Research Council (NERC), as an immediate consequence of which the military-industrial complex typifying British

S. A. Robinson, *Ocean Science and the British Cold War State*,
Palgrave Studies in the History of Science and Technology,
https://doi.org/10.1007/978-3-319-73096-7_7

oceanography since-1945 became more fluid. Once the NIO was no longer linked clearly and unequivocally to military interests in Britain its applications became even more diverse in an attempt to win new patrons. George Deacon's network struggled to adjust to a new funding regime and its new administrators. Maintaining direct relationships with civil servants in the military and their counterparts in the government became even more complex and challenging.

Deacon's personal skills as a mediator could positively impact on his network's expansion only because of the underlying ambitions of his backers in the Admiralty. Since these patrons were no longer interested in promoting Deacon, his role as network builder was less effective. This was especially so because officially the connections between civilian oceanography and military officers, which had allowed for the exchange of information and resources, were technically severed—at least at the most fundamental level of financial support and administrative control. Although less effective than previous associations, informal connections now served an important role in aiding the institution to weather close scientific scrutiny. One outcome of these governance changes was that slowly over time the relationship between the Institute and the state shifted.

This period was a crucial turning point in the history of ocean sciences in Britain (and elsewhere) and in turn was the result of a wider transition away from very stark military-centric interpretations of what was in the national interest (and by extension in science's interest). The preceding years had seen a changed funding regime with the withdrawal of Admiralty funding and the introduction of research councils following a general change in government policy to differentiate 'military' and 'civil' science in Britain. As Deacon fought for money from the NERC and sought patronage from industry and further afield, the NIO's research interests drifted from the military towards the environmental. The second half of the 1960s was a period in which oceanographers had to find a role for their science in a new political landscape. At the centre of this chapter is the story of how British oceanography came to function in a civilianised landscape in interesting ways—which we now largely think of as "environmental". In this change of focus the adaptability of Deacon as a hybrid network mediator, and the support of some of his oldest allies, served to aid the NIO in this transition towards new scientific frontiers.

Redefining the Mission of Oceanography

When the NIO was placed under the NERC in 1965 the institution was formally separated from the military, effectively forming a fully civilian body in both composition and governance.[4] Despite this change in NIO governance, formal separation from the Royal Navy did not result in a cessation of inter-departmental communication.[5] Domestically, even as late as 1970, experts from the Admiralty Underwater Weapons Establishment would still approach NIO experts such as Anthony Laughton for advice on operating and handling oceanographic equipment at sea.[6] At the international level the NIO remained committed to the NATO Sub-committee on Oceanographic Research, or at least up until 1972 when the sub-committee was disbanded.[7] In addition to collaboration between scientists inside and outside the military machine, senior civilian oceanographers including George Deacon were approached to voice opinions on high-level defence research projects such as the NATO MILOC (MILitary OCeanography) surveys.[8] However the purpose of these interactions changed over time. Requests for advice and expertise were now highly specific and linked to challenges being faced by particular programmes. Moreover, oceanographers were being asked for opinions on military matters without necessarily being aware of the specific concerns to naval operations. Being outside the military restricted the NIO scientists' ability to gain the necessary clearance to be a part of highly classified (anti-submarine warfare) projects. Although this had also been the case in Project Corsair, the British attempt to develop its own SOSUS system during the 1950s, it nevertheless proved frustrating to NIO scientists hoping to maintain a two-way flow of information between their organisation and the Royal Navy.

Eventually relations between the military and the NIO cooled. This shift was partly due to the ambiguous status of the civilian posture of the Institute within military–scientific exchanges. They had changed in character since the downfall of British surveillance projects, as outlined in Chap. 4, and as a result of the restructuring of civilian and military oceanographic research at the NATO level.[9] There was therefore a very fine distinction between military and civilian basic research, and this provided precisely the sort of cover that was required to deploy new weapons of surveillance in the oceans. The development of an environmental prediction system to complement and improve SOSUS found little support amongst British military leaders. This resistance originated a questioning not only of such work's scientific merits and basis, but also of the financial challenges faced by the United Kingdom in the years leading up to the 1967 devaluation of sterling. The Royal Navy alleged that the system would bring no benefit but did not have the resources to develop an alternative. Therefore, there was certainly no incentive to engage civilian researchers.

The events of 1963–65, as detailed in the previous chapter, affected Deacon more than anybody else at the NIO. His correspondence and writings at the time suggest that he felt personally aggrieved if not victimised, and he blamed both government and military figures in Whitehall for putting at risk "his" project for British oceanography.[10] In 1965, following these policy negotiations, Deacon complained in a draft article that he circulated to two former National Oceanographic Council chairmen, Sir Frederick Brundrett and Capt. Godfrey French, that he was concerned about developments over the past two years. His concerns concentrated upon what he argued were 'rather nebulous conceptions labeled "military oceanography"'.[11] Oceanography was now being re-configured as an independent pursuit of the Navy, rather than as a coordinated activity involving the NIO oceanographers and Navy officers. According to Deacon those who made new policies in Whitehall had no idea what to prioritise and that had essentially paralysed oceanographic studies in Britain. Despite Deacon's rhetoric, in reality the military had historically allowed quite a lot of lateral freedom in terms of civilian research activities. Rather than being an irritation to oceanographers the ambiguity had been used to full advantage as they were using military funds to develop non-military projects.[12]

The Navy department within the Ministry of Defence (MoD) was concerned about what valuable parts of the NIO research programme they would lose following separation. To quantify this, a Board Committee for Oceanography was formed in 1964 by the naval branch of the MoD and tasked with reassessing the use and role of ocean science within the Royal Navy.[13] The committee was initially chaired by Edmund Irving, the Hydrographer who had been instrumental in the reorientation of Royal Navy science policy between 1959 and 1966. As shown in the previous chapter, this culminated in Deacon's marginalisation from decision-making networks. The initial terms of reference stated that the committee was tasked with 'encouraging and directing the Navy's activities in the field of oceanography'.[14] This was also an exercise in quantifying those segments of the NIO oceanographic programme that were to be transferred to Admiralty research establishments; although ultimately little seems to have been transferred.

Deacon contended that MoD officials considered the NIO 'too academic'. Regardless of whether this was true, basic science had provided the perfect conduit for "military oceanography" at the Institute, whilst allowing money to flow into the NIO. Deacon was then free to re-invest the funds in pursuit of his personal goal of developing British oceanography into a Big Science.[15] Deacon explained these ambitions in "The Future of Marine Science", a draft article that vented his frustrations about the Navy and its mistaken 'loosening [of] its connection with NIO'.[16] Despite the consultations outlined in the previous chapter, he argued that the NIO had not been approached concerning the government's reorganisation of science. Realising that it was not wise to criticise his new governing body, Deacon responded with a statement to say that the NIO was 'satisfied with the argument that an ideal research Council [would] put everything right'.[17] Having previously quarrelled with fisheries scientists in government committees, Deacon was concerned that removing formal ties with the Navy would challenge the position of physical oceanographers and the NIO's programme because there were far fewer non-fisheries scientists to defend such work.

Deacon had deep-seated views about the ideal organisational structure of any research council of which he was to be a part. He feared the activities of the NIO would be judged on its priorities decided by the 'Professors of Zoology that compromise the development commission's fisheries advisory committee' or 'the equally impressive array of professors of geology that run the geology and geophysics subcommittee of the DSIR research grants committee'.[18] This was a very thinly veiled rebuke to Edward Bullard, the Head of the Cambridge Department of Geodesy and Geophyiscs, whom he had clashed with several times over recent years.[19] It is unclear whether Deacon ever intended to publish "The Future of Marine Science", or who his intended audience was. In the covering letter he sent to Brundrett, Deacon wrote 'I may have been a bit more critical of the MOD than is polite, but I can show good reason [sic]'.[20] Both were now out of the military—Deacon under the NERC and Brundrett in official retirement[21]—yet they continued to share ideas and exchange opinions on ocean science and policy, as neither was inclined to act without first obtaining the tacit agreement of the other. They also retained powerful allies, within and outside the government, and liked to think they still held influence—which they did—and that they could direct the policy-making process in their favour—which they largely could not.

Without the safety net of the Admiralty behind him, Deacon saw an opportunity to promote arguments for the use of the oceans for national prosperity, building upon his assertion that Britain should take the initiative and exploit the commercial utility of oceanographic science for the national economy. In this Deacon was supported by Lord Florey, the retiring President of the Royal Society. At his last official function—the Anniversary meeting of the Royal Society on 30 November 1965—Florey took the opportunity to express his frustration at the way in which the question of the future direction and priorities for oceanographic research had been handled by the government during his presidency. He stated:

> Could we make a really great mark in exploring the under-water world… rather than making what at present seem to me to be somewhat pathetically inadequate, though scientifically and technically competent, efforts to enter the same fields as the great rocket powers? Could not a country that mounted the great Challenger expedition foster oceanography and the exploration of the possibilities of the sea, its water, and what is beneath it, so that it would be our national pride and joy?[22]

Florey was echoing remarks made by US President John F. Kennedy in 1963, who argued that 'Our goal is to investigate the world ocean, its boundaries, its properties, its prophecies. It is time to drive back the frontiers of the unknown in the waters which encircle our globe.'[23] Florey's speech was quoted by the *Daily Telegraph* two months later, alongside a detailed report on the "Economic Opportunities in the Oceans".[24] The Royal Society may have ultimately failed to influence Whitehall policy, but a receptive media picked up on the message.[25]

The previous Conservative government of Harold Macmillan had declared that scientific research should be self-governed by scientists with a set research budget, whereas Harold Wilson's newly elected Labour government insisted that nationally funded science should actively support the national economy. With the new government keen to drive forward its new political agenda, the NERC was asked to make the economic case for the orientation of future research. With this in mind Deacon attempted to establish the economic benefit of oceanography to the nation.[26] This followed the first report of the new Council for Scientific Policy, released in May 1966. This report stated that 'it has been represented to us that the capacity of the national economy will not permit the growth of research councils' expenditure at present rates'.[27] The council, which was ultimately part of the machinery of government responsible for the new research councils, stated:

> We have accordingly invited the Research Councils to consider their long-term programmes in order to assess the effect of tapering off of growth rates, particularly in relation to new and scientifically desirable project which might be excluded. We have invited then to develop the justification for their policies, both in terms of intrinsic scientific criteria and in relation to the educational, social and economic benefits.[28]

Deacon was no stranger to justifying the Institute's activities to patrons who had their own mercurial objectives. In response to the shift towards a "rationalisation" of funding he responded as he had before. On previous occasions, when officials in the Admiralty had questioned the military utility of the Institute's activities, Deacon had composed "secret" supplementary annual reports that itemised the more sensitive research.[29] Establishing the Institute's economic pedigree was conducted in a similar manner, although in this case Deacon asked two of his Principal Scientific Officers, David Cartwright of the Physical Division (physical oceanographer) and Laurence Draper of the Electronics & Design & Production Section (the scientific instrument workshop of the Institute), to compose a memo on the commercial value of their particular section's work, rather than compose the memos himself.[30] These internally produced documents were sent to the NERC to be considered by its scientific council.

The NERC circulated minutes on the 'economic case' and 'orientation of future research', having received economic information from several of its institutes. Deacon provided a detailed response that stressed the need for a balanced approach which would bring interdisciplinary research and industry together on major projects. These projects included the redesign and updating of major British ports which at the time was a government priority alongside the emerging market for offshore technologies and industry.[31] Deacon's reply was sent to Raymond Beverton, the secretary of the NERC, who was also a distinguished fishery scientist in his own right. In 1957 Beverton had co-authored, with Sidney Holt, a central text in the overfishing debate, "On the Dynamics of Exploited Fish Populations".[32] Tensions between fisheries scientists and physical oceanographers had been worsening throughout the 1960s and both felt under increasing pressure from reduced government funding. This was exacerbated for fisheries scientists by an increasing sense of crisis in the fishing industry as demand increased and fish stocks declined. Physical oceanographers struggled with their own issues as they shifted their attention from military problems to offshore industries.

Within NATO and at a national level within the NERC, fisheries scientists had repeatedly argued that their science, being tied closely to the fishing industry, should be prioritised on the basis of its applicability to economic activity. Friction between these different groups of scientists within the NERC regularly resurfaced. In his correspondence with the NERC Deacon argued that his institution's work was suffering an injustice through an "elementary" portrayal of the state of the field by other ocean scientists whose work was primarily concerned with biological questions.[33] Deacon attempted to re-cast military research that had been undertaken just a few years earlier with a perceived defence purpose, as civilian. He framed this same work as beneficial to both fisheries research and fishermen who, he argued, needed 'synoptic information and long-term forecasts'.[34] This had been the primary focus of physical oceanographers since the International Geophysical Year (1957–88), and reflected the research agenda that they continued to pursue. Deacon essentially used his skills as a network builder to reconfigure the web of relations that he had managed up until that point. However, as has been shown, Deacon's skills could positively impact on the network's expansion only because of the underlying ambitions of his backers within the Admiralty. With these patrons no longer interested in promoting Deacon, his role as network builder was less effective, especially amongst fellow scientists.

Furthermore, encroaching on the territory of fisheries scientists in this manner engendered distrust amongst those scientists who, thanks to their years working on such questions, felt well placed to articulate what fishermen and fisheries scientists actually needed. It also showed an underestimation at the NIO of the importance that biological and chemical parameters had, compared to physical variables, in addressing specific issues. Unlike the initial memoranda prepared by Cartwright and Draper on the economic basis of current research, Deacon's "edited" version was heavily influenced by how he felt government policy was likely to develop. Additionally in his revisions he argued that unnamed companies were showing an interest in ocean science and its adaptability to their commercial activities, offering the potential for financial patronage in the future.[35]

Deacon's letter to Beverton made use of the fact that large companies as well as scientists, governments, and major scientific institutions were reassessing the potential of the oceans. He intended the NIO to be poised to reap any potential profits from this movement towards the commercialisation of ocean resources.[36] In 1965 a Commercial Oceanology Study Group was formed, bringing together six major British corporations: British Petroleum (oil), Unilever (consumer cleaning and food products), Richard Costain (construction),[37] Imperial Chemical Industries (chemicals), Rio-Tinto Zinc Corporations (mining), and Hawker Siddeley Group (aerospace/engineering/defence contractor).[38] Together they agreed to sponsor one research project at the NIO to study the long-term possibilities of sea exploitation.[39] This re-imagining of the ocean as a bountiful new economic frontier, to be explored and ultimately exploited using the tools of science and engineering, was given a great boost when BP's rig *Sea Gem* found traces of gas in the British sector of the North Sea in September 1965.[40] Whilst the commercial interest in the exploitation of the oceans was growing in Britain, defence research continued—although it should be noted that military and commercial oceanography were seldom discussed together. In a 1966 *Science* article that appeared alongside a report on NERC research proposals for oceanography, it was noted that 'an interesting feature of these arguments is that they hardly mention the defence applications of marine science'.[41] With the Labour government keen to link science and technology to industry, and to draw traditional science out of its perceived ivory tower, defence was increasingly overshadowed by attempts to suggest that their approach and policies could bring about economic recovery. The justification for ocean science had shifted from an orientation based on military utility to one of economic merit.

Crafting Oceanology

Beginning around 1966, the term "oceanology" began to circulate in science publications, reflecting the new commercial drive to exploit the resources of the oceans. The neologism carefully distinguished activities that were commercial in nature from oceanography, which was considered purely scientific.[42] Oceanology had a very broad definition at the time. Robert Barton, in his 1970 book *Oceanology Today*, explained that oceanology was a new word that had yet to be fully defined. Barton claimed that oceanology was an attempt to bring together the 'hitherto fragmented, multidisciplinary approach to marine exploitation, with scientific investigation directed to economic ends rather than to the gathering of knowledge for knowledge's sake'.[43] Within these deliberately vague definitions of the term, "oceanology" seemed part applied research and part capitalist crusade in the form of a movement of individuals with the shared objective of opening up the oceans to exploitation. This was in much the same way that land resources were being exploited as worldwide consumption of resources escalated.[44] Behind all this rhetoric of science and industry in partnership for the benefit of mankind were simple hard economic realities tied in part to the continuing concerns of Cold War resource security combined with increasing global consumption driving up prices and demand.[45]

Constructing the term "oceanology" was an objective of ocean scientists, including the likes of Deacon and Brundrett, rather than the companies that were already operating in British waters searching for offshore oil and gas deposits. The blending of marine exploitation with scientific investigation was designed to benefit the scientists in particular as experts in commercial ventures. The interest of offshore companies in broader understandings of the oceans was limited, their objectives were driven by profits, and their goal was to find oil deposits. The sincerity of any professed interest beyond this goal was dubious. Therefore, the interest of industry had to be courted, and Brundrett and Deacon felt they needed to show their value to these new wealthy companies.

To this end Brundrett turned to the new market of semi-popular science publishing.[46] He continued to be both a scientific adviser to government through the White Fish Authority and an advocate of marine industries through publishing.[47] In becoming both civil servant and lobbyist, he was stretching the definition of an independent civil service, a duality that had been partly responsible for his downfall at the MoD in 1959. His influence had waned since Solly Zuckerman replaced him as Chief Scientific Adviser at the MoD, and he could no longer directly influence the likelihood of patronage from the government, though as a powerful protagonist using a network of contacts and his own personal reputation he could give voice to British interests in marine science.

Oceanology International and *Hydrospace* first appeared in 1967 as bimonthly periodicals in the United States and United Kingdom respectively. They were intended to chronicle developments in marine affairs, both industrial and scientific, and *Oceanology International* sold itself as the 'spokesman for ocean science and technology'.[48] Each exposed the latest attempt by oceanographers to demonstrate that their science was worthy of increased investment and prestige similar to the funding, media coverage, and political support afforded to space research at the time.[49] The appearance of publications rather than the creation of a new international oceanographic committee, council, or other body reflected a shift in the governance of oceanography by practitioners. It also gave the appearance of oceanology being itself a new science, independent of the complex politics of international oceanography.

International committees, bodies, and institutions required the tacit support of governments and military patrons to establish and ultimately fund them. International bodies were expensive, complicated, and capable of coordinating the expenditure of vast amounts of participants' national research budgets for limited national economic benefit, while companies worked under much closer financial scrutiny and were driven by profits.[50] These new publications were funded by extensive advertising from companies speculating in marine industry, alongside companies—such as Plessey and Vickers—who had previously supplied scientific equipment to the Royal Navy and commercial operators. This funding gave such publications a degree of financial independence that was difficult to achieve through other means. Plessey had previously been given the right to build the NATO-developed ocean current meter in exchange for royalties that had been used to fund NATO research. Although these advertisements often showed drawings of the company's wares in action on the high seas, it was claimed by the head of the workshops at the NIO that most of the products advertised in glossy US magazines were merely concepts and that the companies were ultimately searching out patrons themselves in order to bring money in to develop these devices.[51] The imagined market for marine technology was to be stimulated by the glossy outputs of advertising agencies, but its drivers off the page had yet to emerge, because of the vast financial outlay ocean exploitation required.

Oceanology International and *Hydrospace* attempted therefore to sell ocean science and technology as new and modern, which was not only reflected in their design but in their language and titles. For example, "hydrospace" was a word coined by science-fiction writers to reflect inner space as opposed to outer space, and writer Arthur C. Clarke was a member of the editorial board of the US publication *Oceanology International*.[52] Yet it was a British publication that chose this term as its title, invoking the importance of ocean space as a topic for discussion rather than merely for its exploitation. In November 1967, in the editorial for the first issue of *Hydrospace*, Brundrett argued that the ocean was a void, or a frontier where exploitation offered rewards for the nation that pioneered such developments.

> Great Britain is well fitted by its geography, through its scientists and engineers – and, most important, through the history and outlook of its people – to play a major part in shaping the future role of the seas and oceans of the world; success is assured if there is proper direction and scale of effort![53]

In this case Brundrett and the editor of *Hydrospace* Robert Barton were responding to a growing interest in Britain in the exploitation of the oceans.[54] Yet, Deacon was not entirely convinced by oceanology, and no other earth science discipline had devised a second disciplinary label for its applied research. While he obviously shared with Brundrett the wish to find new connections in the business world that would allow ocean studies to thrive, he was quoted as stating that the word sounded 'like some rather obscure cult, like astrology, rather than conveying the true gravitas of what was really being undertaken "ocean engineering"'.[55] This sentiment reflected Deacon's difficulty in adjusting to the profound changes that were occurring in the seas surrounding the United Kingdom as a consequence of a number of recent events, such as the search for offshore oil and gas, the debates surrounding the conservation of fish stocks, and marine pollution featuring as a national and international concern.

Whilst selling ocean science and technology, as well as the oceans themselves to potential investors, these publications also reflected not only the company's attempts to sell their products, but the efforts of oceanographers to sell their expertise. The NIO attempted to market technologies and scientific knowledge gathered and developed whilst its oceanographers were still involved in collaboration with the military. Three devices typified this new drive towards commercialisation during this period. Each had been developed at the NIO, but never made operational before 1965. Now the thermistor chain, the Geological Long-Range Inclined ASDIC (GLORIA), and the Institute's geological maps of the seas surrounding the United Kingdom became key spin-off technologies of earlier military work (see Fig. 7.1). All of these became potential products for the emerging industries concerned with ocean exploitation.

The thermistor chain had been developed both to make data collection more efficient and to open up the deep ocean primarily for military purposes, but it was just as useful to civilian oceanographers.[56] Before the 1960s scientists had relied on bringing the deep sea to the surface in

Fig. 7.1 Geological Long-Range Inclined ASDIC (GLORIA) Trials team aboard the RSS *Discovery*. *L to R Seated Front* Ray Peters, Dick Dobson, Stuart Bicknell. *Middle* John Swallow, Ship's Officer, Norman Smith, Harry Moreton (bosun), Dick Burt (netman), Ship's Officer, Capt. Geoff Howe, unknown, Stuart Rusby, Mike Somers, Brian McCartney. *Back* Brian Barrow, Vince Lawford, Keith Tipping, Stuart Willis, Roger Edge, Percy Woods

Nansen bottles, nets, and dredges, or had measured the ocean's dynamics with bathythermographs and current meters.[57] The development of deep-sea diving and underwater television had begun to visually open up the deep ocean during the 1950s.[58] The problem, however, was that oceanographic vessels could only dangle one device at a time from their winches and often operating different types of equipment required the ship to be handled in different ways.[59] Beyond this, taking one measurement in one place made oceanographic work expensive and made synoptic studies impossible without collaboration between vessels. This was the driving force behind the international synoptic studies of the world's oceans that occurred in the 1950s and 1960s such as the IGY, IIOE, and the NATO Science Committee's activities. However the thermistor chain allowed one vessel to undertake work that would have previously required multiple ships and lots of preparation. A thermistor chain was a series of underwater temperature measurement gauges attached to one long cable; these enabled simultaneous measurement of temperature through a tall water column, allowing scientists to observe the thermal layering of deep ocean water.

GLORIA was one of the most powerful side-scan sonar that existed at the time and its multifaceted applications could generate contracts. The NIO had a lead on the world for the first time and took every opportunity to promote their device on the international stage. The instrument was capable of operating at depths of 21,000 feet and used 144 transducers in a 32-foot-long yellow case which weighed in at 6.5 tons. Initially developed by the biologists at the NIO, GLORIA was an instrument that could be towed behind the *Discovery* and could use echo sounders to track the movement of marine life, helping the military authorities to understand sources of noise in the detection of enemy vessels. GLORIA, however, was one of the programmes that Deacon had agreed not to prioritise, given the existence of the other surveillance programmes. However, early in the development process it became apparent to John Swallow and Anthony Laughton that the device's sensors, if orientated slightly differently, would enable detailed topographical surveys of the sea floor. This would build a three-dimensional picture, the kind of valuable picture that a lead weight dropped on a line, or the explosive geological profiling method developed by Edward Bullard's team at Cambridge, could not produce. Now Deacon and his assistants understood that they had a powerful device that could be devoted to a variety of "oceanological" applications.

Although popular science writer Robert Barton claimed that GLORIA could produce images much like aerial reconnaissance, this was an exaggeration, and even the grainy examples he reproduced as an illustration required extensive labelling to be comprehensible.[60] Once deployed the instrument would require little handling from the ship's company and would produce data on a print-out plot, which could be collected in the ship and interpreted later. The most immediate application of powerful side-scan sonar was to replace the existing practice of using explosives to gain a seismic picture of the sea bed. This was extremely expensive, using up to ten tons of explosives a day at a cost of US$100,000 a month.[61] Beyond this the potential applications for side-scan sonar were diverse. Early devices had been used to search for a hydrogen bomb lost by the US Air Force in an accident off Palomares, Spain in 1966, although it subsequently became apparent that operators found it difficult to distinguish between the bomb, fish, and other debris; instead the US Navy's deep-sea submersible *Aluminaut* was used to recover the device.[62] Non-military applications included underwater surveying for the placement of drilling rigs and pipelines; eventually there would be a multiplicity of applications for these devices including scouring the deep sea for shipwrecks. However one of the most profitable uses of GLORIA for the NIO came from the US government, but only in the 1980s.[63]

Side-scan sonar was one of a group of exciting new technologies that Barton and other commentators claimed would open up the ocean depths to humanity.[64] Essentially these devices turned the RRS *Discovery* into an instrument platform, capable of doing the work of a fleet of oceanographic vessels.[65] The motivation behind these improvements in ship efficiency lay in the poor economic situation of the country and the desire to remain in the front rank of oceanographic nations.[66] The secretary of the NERC, Raymond Beverton, told a reporter from *Science* that 'the prime need at this stage is to keep up with the rapidly developing instrumentation. Oceanography is moving toward "Big Science" and these devices represented the results of this thinking.'[67] Whilst other nations, particularly the USA, had been producing bathyscaphs in order to take humans to the deepest parts of the ocean, the British seemingly by chance had taken a very different approach, deploying instruments rather than humans to the ocean depths.[68]

In Britain the oil and gas deposits in the North Sea also heralded great potential for improving national resource security.[69] This prompted the third ocean technology marketed by the NIO post 1965—its ocean data sets. As head of the Geology Division at the Institute it was part of Anthony Laughton's role to foster these links with industry. The oil industry's interest went beyond the geological profile of the sea bed. The North Sea was a much deeper sea and had a different climatic environment than most offshore oil companies, used to working in the Gulf of Mexico, had worked with before. Hurricane-force winds were unusual in the North Sea but continual battering by lower-grade sea conditions led to rigs collapsing because of fatigue.[70] The NIO had gathered information on the wave conditions around the British coast since before its foundation through the work of the W group at Perranporth Beach in 1944. This knowledge was now used to predict the maximum expected wave heights for a given area, so rigs could be constructed as economically as possible whilst remaining "safe"; additionally, sea-state predictions gave engineers an indication of the required specifications of metal materials used in rig construction. Monitoring ocean conditions became a priority and for oil companies especially after 1967, when the *Torrey Canyon* disaster, discussed below, focused concern on the ecological impacts of the spillage of crude oil into the oceans.

By the late 1960s there was growing political and public pressure to separate science from the military, particularly in the United States.[71] This was in addition to a move towards the peaceful use of oceanographic science and calls for the greater exploitation of what appeared to be abundant marine resources.[72] In this period Deacon's marginalisation within the civil service became apparent in the handling of the *Torrey Canyon* disaster. Deacon's personal scientific standing allowed him to present himself as an expert who knew better than his peers, presenting himself and his colleagues as having expertise superior to his rivals, particularly fisheries scientists, in an attempt to validate yet another avenue of expansion for the NIO. This caused confrontation at NATO within the oceanography subcommittee.[73] Deacon began to look closely at the expansion of the NIO into a second building at Wormley, and attempted to build links with universities to set the NIO on a higher academic plane than the applied fisheries scientists. However, this usual refrain of infighting between fisheries scientists and physical oceanographers was challenged by a series of environmental "mishaps" on the oceans.

A series of marine environmental accidents gave rise to concerns as to the impact of mankind's exploitative practices on the ecology of the sea. Oceanographers in Britain contributed to this new environmental discourse but they were neither advocates of environmentalism nor did they seek to defend the companies whose actions were being lambasted, by the former, as reckless. As Hamblin has demonstrated in the case of the dumping of radioactive waste at sea, the question for oceanography became one of determining the capacity of the ocean to cope with man-made pollution.[74] The aftermath of the *Torrey Canyon* oil spill demonstrated both the lack of access the NIO had to those in power, and also the continuing ability of fisheries scientists to influence those in power, even when the scientists knew their only solution to the catastrophe would be even more damaging. This episode is symbolic of the post-1965 malaise in the relationship between the NIO and the higher offices of state, which had hitherto been much closer. However, it can also be argued that this event was one which exposed not so much new avenues of research, current and drift research being well established, rather highlighting new applications for existing research if it could only be communicated to a more general audience.

These divisions were brought into the open when the oil tanker *Torrey Canyon* ran aground in 1967 releasing 119,000 tons of crude oil. The Ministry of Agriculture, Fisheries and Food's (MAFF) official report published a year later in 1968 provided a simple, analytical appraisal of events: '30,000 tons of oil were released on to the sea at the time of the wreck...20,000 tons were lost during the next seven days ...the vessel broke her back, releasing a further 50,000 tons. On 28–30 March the "Torrey Canyon" was bombed...20,000 tons of oil was destroyed by burning; negligible quantities of oil remained in the tanker after 30th March.'[75] The report did not mention that before bombing the wreck with napalm, large quantities of highly toxic detergent had been deployed in an attempt to dissolve the oil. This caused greater damage to marine life than the oil had in the first place. MAFF (authors of the report) had known of the toxic effects of detergents since 1965, and this was handily included as an appendix to the 1968 *Torrey Canyon* report.[76] The suggestion of bombing the tanker had been made by the new Cabinet Office's Chief Scientific Adviser, Sir Solly Zuckerman,[77] during a Cabinet meeting on Sunday 26 March which was held at Culdrose Royal Navy Air Station, Cornwall, near the disaster so as to ensure maximum publicity.[78] All of this was designed to defend institutional reputations, but the scientific response to the disaster had been, quite evidently, as damaging as the sinking itself. Neither MAFF nor Zuckerman wanted to take any blame.

Zuckerman himself wrote a report, published by the government, which highlighted the use of detergents and downplayed the bombing of the oil on the surface. The one official voice that remained silent, at least within government publications, was the NIO. Deacon was as quick to respond to the disaster as fisheries scientists had been and he felt that physical oceanographers—himself especially—had not been consulted. Deacon felt that the premise that the oil would disperse quickly was ultimately flawed and was a result of a general lack of understanding of the physical properties of the sea, especially by fisheries scientists.[79] This was not just a battle for ocean politics but also an opportunity to find new patrons by exploiting the disaster. It seemed likely that with an increasing tonnage of oil-carrying vessels being launched each year, such a disaster would occur again in the future. However, having failed to find an avenue at the time to advise decision makers in government, Deacon made the disaster the subject of his address to the British Association for the Advancement of Science meeting in 1968.[80]

From this platform Deacon was highly critical of the efforts of fisheries scientists and the advice they had given, and outlined the existing research on sea surface drift (oil floats on the surface), which was affected as much by the winds as by tides and currents. It seemed unbelievable that the episode had turned into such a debacle.[81] These sentiments echoed the comments made by J.N. Carruthers that the solutions to the problems associated with British ASDIC operations during the Second World War had been openly accessible in pre-war Norwegian fisheries journals.[82] Carruthers' point, like Deacon's, was that because of an emphasis on new research rather than looking back to past work, mistakes were easily made and past experiences not explored. In this case the research on sea drift had been undertaken in Germany in the late 1950s but the lessons were not being heeded or communicated.[83]

Solly Zuckerman's report into the *Torrey Canyon* disaster had stated that there was a lack of research on oil dispersion, which was being rectified by the NIO as a lesson learned from the incident. Conversely, Deacon claimed this research already existed and that research into air–sea interaction with oil slicks was already well developed.[84] The poisoning of marine life by crude oil came as no surprise, but the fact that detergents used as dispersants for the oil had also caused damage has resonated ever since as a warning of the need to carefully mitigate environmental damage.[85] The process by which the physical action of the sea contaminated marine life was a chemical one. Chemical oceanographer Max Bulmer of the Woods Hole Oceanographic Institute (WHOI) in the USA produced a report entitled 'Oil Pollution of the Ocean'.[86] The report was particularly damning of what it saw as the reckless nature of oil shipping companies, stating 'all pollution is the almost inevitable consequence of the dependence on a largely oil-based technology'. Woods Hole scientists were compelled into action following an oil spill off the coast of New England in September 1969, practically on the doorstep of their institution. These events brought home to scientists the biological catastrophe of oil pollution. A particularly emotive line from a subsequent article by two WHOI marine biologists reflected that 'nature works for man and man works against it'.[87]

Although the *Torrey Canyon* episode was the catalyst for what may be termed British and American marine scientific environmentalism, the NIO's unblemished reputation, insofar as it could not be blamed in any way for the failure of the clean-up operations, presented an opportunity for Deacon and a new avenue of research for the Institute. Once again he viewed environmentalism as a possibility to expand his network and find new sponsors.

Concern for the protection of the marine environment opened up new research opportunities for the NIO and earth scientists elsewhere in Britain. Although environmental historians often point to Rachel Carson's *Silent Spring* as the genesis of scientific concerns about pollution of the environment, the mishandling of the *Torrey Canyon* incident had a more profound impact on the scientific concerns of oceanographers.[88] In Britain the *Marine Pollution Bulletin* (*MPB*) was published as just such a cross-over publication by Macmillan and edited by Professor Robert Clark of the Department of Zoology, University of Newcastle. *MPB* was first published as a mimeograph in 1968 but the rapid growth of interest in the field attracted the interest of the publishing house Macmillan, and then Robert Maxwell's Pergamon Press (in 1975).[89] In the very first issue Clark emphasised that this was to be 'a monthly information bulletin, not a scientific journal'; the contributors would be 'scientifically active' but remain anonymous. Contributions to *MPB* reflected a growing professional concern amongst scientists of the effects of pollution on the oceans. The contents of this basic publication covered topics such as oil spills, fish kills, unusual seabird mortalities, advances in anti-pollution techniques, and useful further sources of information.[90] Part scientific journal, part newspaper, and part ecology opinion piece, the mimeograph morphed into a professionally produced journal when Macmillan started to print and distribute it from January 1970.

Alongside this publication, which to an extent bridged scientific concerns and political activism, there was also a shift towards new peer-reviewed scientific publications dedicated to pollution and the marine environment. Beyond more specialist literature, in 1970 the popular environmental magazine *The Ecologist* was first published in Britain. This year had been designated the European Conservation Year, in which there was to be a concerted effort to combat rising atmospheric pollution. Before the publication of *The Ecologist*, much of the scientific/environmental debate had been conducted in other more mainstream popular science publications such as *New Scientist*. However the emergence of this publication demonstrated that environmental concern was a large enough single issue to sustain the production costs. In August 1970 an article appeared in *The Ecologist* linking the "population bomb" to mineral resource exhaustion and arguing that the exploitative industries which oceanology encompassed heralded no less than the potential end of the world.[91] The question of marine pollution was the focus of the first half of 1971 with articles on the "Prevention of Marine Pollution" (January),

"Cousteau on ocean pollution" (February), "Can the Sea Survive?" (March), "Aluminium and Anglesey" (June), and on the litigation following the *Torrey Canyon* disaster (July).[92] David George, marine ecologist at the Natural History Museum, authored "Can the Sea Survive?", but the other articles were all written by staff or freelance writers. This was the beginning of an environmental movement in Britain based on ecological thinking, which eventually spawned the Friends of the Earth organisation. However the existing historiography of the environmental movement in Britain is limited, and the commitment to, and relationship with, these organisations by members of the NIO is somewhat dubious.[93]

The societal expectations of science, and specifically ocean science, had shifted during the late 1960s. It was more than simply a matter of science being used for military purposes; science could be and was being mobilised for environmental concerns. This was ultimately realised through the NATO Committee on the Challenges of Modern Society, established by President Nixon in 1969.[94] As Hamblin has shown, there was a degree of European scepticism towards investing money, through the Atlantic alliance, in environmental research. And despite accusations that the USA was attempting to influence the upcoming UN Conference on the Human Environment, scheduled to take place in Stockholm in the summer of 1972, the reality was that the mission of Cold War science was shifting towards new goals. Ron Doel has argued that in the 1960s there emerged two distinct sciences of the environment, one driven by military-operational needs and the other focused on ecology.[95] However in the case of British oceanography the two sides of this science came from the same group of scientists responding to events of which they were as much a part as the activists who used them for political purposes. The result of this was a duplicity in, and diversification of, the use and application of the tools of oceanographers, who no longer felt they could rely on the support of their previous military patrons, towards new missions.

The Military Twist to Environmental Research

While pursuing environmental research, Deacon continued to promote defence research. In 1968 he was once again in communication with the Ministry of Defence regarding the proposed NATO MILOC surveys, to be carried out by the Royal Navy off the Norwegian coast in 1969. The NIO had been excluded from these surveys, as it had been from British military oceanography/hydrography since 1965. In a letter from Bill Kelly

(Director of Naval Physical Research) to Deacon, the former outlined the failings of the NATO oceanography centre at SACLANT to successfully analyse the previous survey's data, and invited the NIO to help.[96] Considering Deacon's recent public statements, his private response was positively gushing. 'We get a lot of help from the Navy and would like to do all we can in return.' Although he admitted he had not followed the recent developments in the MILOC programme, he clearly knew enough to state that he was aware that the next programme would be concerned with synoptic observations of the variability of the temperature gradients in the main thermocline through space and time. He went so far as to state that the NIO was interested in very similar questions. Echoing much earlier arguments, he ended 'it would be a good thing for marine science as a whole if we could combine forces'.[97] As a postscript to this letter Deacon added 'we are very keen to help if we can'. This was not the only softening of Deacon's rhetoric that seemed to be occurring. Deacon circulated a minute amongst the members of the Royal Society's British National Committee for Oceanographic Research (BNCOR) acknowledging that 'biology like physics seems to need new facilities'.[98] This was very different to his earlier advocacy of the need for investment in physics because of the imbalance that favoured biological and zoological research. Deacon seemed to have finally realised that whilst scientists often stayed in post for a long time, senior military officers and others within government moved on with a convenient regularity. It was a reality that had worked both for and against Deacon over the years.

Given the new opportunities that began to present themselves, Deacon was cautious. Ultimately the NIO withdrew its offer of helping the MILOC programme once it became apparent that the offer would attract no additional funds or facilities. Deacon felt that the analysis would take his scientists away from other valuable work that the NIO was pursuing and for which the military had shown little support. Furthermore, the years during which he was left out of military matters had forced him to look to new avenues for research. Even if the NIO had kept an eye on military developments and had voiced opinions when the opportunity arose, they had also secured independent funding through the NERC. This funding meant that the NIO had the luxury of saying no, and when the military came back they found the scientists willing to act as external experts, not internal lackeys. The result was a much more independent institution and a much more scientifically diverse one in terms of its ability to adapt its programme to new societal concerns about the environment.

With the growth of concern over marine pollution, the marine environment, and humanity's destruction of the oceans, the focus of Britain's primary institute of oceanography shifted to an extent in line with wider societal, political, and cultural changes of the period. On the one hand this was a broadening out of the applications of basic research, on the other hand it was an opportunity for new non-military avenues for research. New patronage networks, however, failed to completely provide the security that would have come if the income they generated had risen to a level equal to the operating costs of the Institute. Governments' need to understand the potential impacts of business and commerce on the seas returned oceanographers to a prominent place in maritime politics. Deacon and others may not have had the same direct contact with figures in Whitehall as they had done previously, but discussions were no longer confined to the halls of power. Debates in the popular press, at UN meetings, and with academics from non-scientific disciplines propelled oceanographers towards a fully independent status. Most of the popular scientific "glory" during the period may have been in space but the debates surrounding the health of the ocean were seen in environmental discourse to reflect the health of the planet as a whole.

This propelled oceanography beyond its traditional military applications. Interest in ocean science by the wider public in both Britain and the United States led to the rise in university-level courses in oceanography for which Deacon and others had been arguing—as shown in previous chapters—for the past fifteen years. Deacon was hoping that he could reconfigure the same expertise of military oceanography in different environments, but the results of this were mixed. Commercial and industrial concerns may have reflected the political direction to which science was being put by the Labour government, but Deacon failed to win the argument as to the value of the NIO's research when environmental matters were at stake, as witnessed by the use of his long-term nemesis, fisheries scientists, as advisers in the *Torrey Canyon* disaster. In this new post-military era in oceanography, the networks which Deacon had relied upon previously were replaced with short-term connections built on flimsy relations that often disappeared abruptly or were relevant only for the length of the project or the collaboration. Nevertheless, the concept of networks and the importance Deacon attached to them resulted in a ceaseless attempt to foster, create, and maintain them even under shifting circumstances.

Notes

1. Robert Cockburn, "Science, Defence and Society," *Journal of the Royal Society of Arts*, 115:5131 (June, 1967): 548.
2. On the shifting relationship between science and the military during the late 1960s, see John Krige, "The Peaceful Atom as Political Weapon, Euratom and American Foreign Policy in the late 1950s," *Historical Studies in the Natural Sciences,* 38:1 (2008): 5–44; Simone Turchetti, Simon Naylor, Katrina Dean, Martin Siegert, "On thick ice: scientific internationalism and Antarctic affairs, 1957–1980," *History & Technology*, 24:4 (2008): 351–376; Mark Solovey, "Project Camelot and the 1960s Epistemological Revolution: Rethinking the Politics-patronage-social science nexus," *Social Studies of Science,* 31:2 (2001): 171–206; Holger Nehring, "National Internationalists: Britain and West German Protests against Nuclear Weapons, the politics of Transnational Communications and the Social History of the Cold War, 1957–1964," *Contemporary European History,* 14:4 (2005): 559–582.
3. Peder Roberts, '"What has all this got to do with science?' The Rhetoric of Scientific devotion in British Government plans for the International geophysical year," *Proceedings of the 3rd Workshop of the SCAR Action Group on the History of Antarctic Research,* https://kb.osu.edu/dspace/bitstream/handle/1811/BPRC_Tech_Rept_2011-01.pdf accessed 19/2/2014.
4. 'Research Reshaped: Commons Debate on Technology', *The Guardian,* 12 December 1964; see also Chap. 5 of this book.
5. As shown by Simone Turchetti, Deacon continued to partake in the NATO oceanography sub-committee on behalf of British oceanographers. Simone Turchetti, "Sword, Shield and Buoys: A History of the NATO Sub-Committee on Oceanographic Research, 1959–1973", *Centaurus* 54:3 (2012) 226.
6. Laughton was said "[to have] had the most experience of towing things near the bottom with pingers attached and is interested in high resolution bathymetry, would be able to discuss the problems", from letter from Brian S. McCartney to M.J. Daintith (Admiralty Underwater Weapons Establishment), 30 December 1970, *Secret,* GERD Papers D5/3, NOC Library (Southampton).
7. See footnote 6 above.
8. MILOC (Military Oceanography) surveys were conducted under the auspices of NATO Defence Research Directors; letter from H.W.K. "Bill" Kelly (Director, Naval Physical Research) to Deacon, 26 August 1968, GERD Papers, D5/3, NOC Library (Southampton).

9. See John Krige, "NATO and the strengthening of western science in the post-Sputnik era," *Minerva* 38:1 (2000): 81–108; Simone Turchetti, "Sword, Shield and Buoys: A History of the NATO Sub-Committee on Oceanographic Research, 1959–1973," *Centaurus* 54:2 (2012): 205–231.
10. See correspondence between Deacon and Captain Godfrey French (R.N. Hydrography Office) 1964–5, GERD Papers, D8/1, NOC Library (Southampton).
11. 'The Future of Marine Science', drafted by George Deacon, undated [but early 1965], GERD Papers, D1/10, NOC Library (Southampton).
12. Particularly for surveillance or sensitive surveying work; see Chap. 4 of this book.
13. Board committees functioned much like sub-committees or working parties, and reported to the board of the Admiralty, which was chaired by the First Sea Lord, Sir David Luce (ex-submarine commander).
14. Ministry of Defence Office Memorandum (Navy Department), 25 November 1964, N/RDF19/64, DEFE 69/485,TNA (London).
15. 'The Future of Marine Science', drafted by George Deacon, undated [but early 1965], GERD Papers, D1/10, NOC Library (Southampton).
16. Ibid.
17. Ibid.
18. Ibid.
19. At the time Bullard was building ever stronger relationships with American colleagues, such as Roger Revelle, whilst Deacon seemed to merely exacerbate them. See letter from Revelle to Bullard, 25 February 1964, ECBP, C.18, CAC.
20. Letter from Deacon to Fredrick Brundrett (Scientific Adviser to the Civil Service Commission), 9 February 1965, GERD Papers, D1/10, NOC Library (Southampton).
21. Despite official retirement Frederick Brundrett continued working for the Civil Service Commission, and as an overseer of the administration of the civil service across Whitehall. For further information see the extensive correspondence between the two contained within the GERD papers at Southampton.
22. "Address of the President, Lord Florey, O.M., at the Anniversary Meeting, 30 November 1965," *Proceedings of the Royal Society of London. Series B, Biological Sciences*, 163:993, (Jan. 18, 1966): 425–434.
23. JFK first used the term "new frontier" in his acceptance speech for the Democratic nomination for the 1960 presidential election. See John F. Kennedy, 'Address of Senator John F. Kennedy Accepting the Democratic Party Nomination for the Presidency of the United States, 15 July, 1960.' Available at: http://www.jfklibrary.org/Research/Ready- Reference/

JFK-Speeches/. Quote: 'We stand at the edge of a New Frontier – the frontier of unfulfilled hopes and dreams, a frontier of unknown opportunities and beliefs in peril. Beyond that frontier are uncharted areas of science and space, unsolved problems of peace and war, unconquered problems of ignorance and prejudice, unanswered questions of poverty and surplus.'

24. A. Michaelis, "Neglected Riches of the Deep," *Daily Telegraph,* 22 January 1966.
25. N.H. Gaber and D.F. Reynolds, "Economic Opportunities in the Oceans," *Battelle Technical Review,* 14:13 (December 1965): 5–11.
26. Letter from Deacon to R.J.H. Beverton, 27 April 1966, GERD Papers, D8/1, NOC Library (Southampton).
27. Quoted in Hilary Rose and Stephen Rose, *Science and Society* (London: Allen Lane, 1969): 108.
28. Ibid., 109.
29. 'Supplement to Annual Report for Year Ending 31 March 1950: Application of the Institute's work to Defence Problems.' This was very similar to the list provided to the Admiralty in 1947, *Secret,* GERD Papers, D1/3, NOC Library (Southampton).
30. Cartwright, D.E., 'Some work of the National Institute of Oceanography of Commercial Value to the Nation', (14 February 1966); Draper, L., 'Notes on Engineering Benefits' (18 February 1966), GERD Papers D8/1, NOC Library (Southampton).
31. Richard Goss, "British Ports Policies Since 1945," *Journal of Transport Economics and Policy,* 32:1 (1998): 58.
32. Letter Deacon to Raymond Beverton (Secretary of NERC), 27 April 1966, GERD Papers, D8/1, NOC Library (Southampton). Beverton had co-authored with Sydney Holt *On the Dynamics of Exploitated Fish Populations* (Ministry of Agriculture, Fisheries and Food, 1957), an influential early contribution to the overfishing debate.
33. Ibid.
34. Ibid.
35. Ibid.
36. In an article for *Science* this was claimed to be due to 'Britain [being] increasingly aware of its perilous economic situations'; Victor K. McElheny, "Oceanography in Britain: Significant New Support," *Science* 153:3737 (1966): 727–8.
37. Would go on to build the Thames Barrier (1984) and was part of the consortium which built the Channel Tunnel completed in 1994.
38. Robert Barton, *Oceanology Today* (London: Aldus, 1970): 23.
39. "Six firms sponsor sea study," *Guardian,* 17 July 1965.
40. A. Bambridge, "BP Steps on the gas," *The Observer,* 26 September 1965; for more on the Cold War and North Sea oil, see Roberto Cantoni and

Leucha Veneer, "Underground and Underwater: Oil Security in France and Britain during the Cold War," in *The Surveillance Imperative: Geosciences during the Cold War and Beyond*, ed. Simone Turchetti and Peder Roberts (London: Palgrave, 2014) 55–58.

41. Victor K. McElheny, "Oceanography in Britain: Significant New Support," *Science* (12 August 1966): 727–8.
42. On the etymology of the term oceanology, see "oceanology. n." OED Online. March 2014. Oxford University Press. http://www.oed.com/view/Entry/130221 (accessed 1/5/14); for an earlier discussion of the differentiation between oceanography and oceanology, see James N. Carruthers, "'A Plea for oceanology," *Deep Sea Research* 2:3 (April, 1955): 247.
43. Barton, *Oceanology Today* 9–10.
44. Richard P. Tucker, "The International Environmental Movement and the Cold War," in *The Oxford Handbook of the Cold War*, ed. Richard H. Immerman and Petra Goedde (Oxford: Oxford University Press, 2013): 565–583; Kai Hünemörder, "Environmental Crisis and Soft Politics: Détente and the Global Environment, 1968–1975," in *Environmental Histories of the Cold War*, ed. J.R. McNeill and Corinna R. Unger (Cambridge: Cambridge University Press, 2010): 257–278.
45. Cantoni and Veener, "Underground and Underwater," 45.
46. See Iain Stevenson, *Book Makers: British Publishing in the Twentieth Century* (London: British Library, 2010): 141–144. One of the major players in this market was Robert Maxwell's Pergamon Press, but there were also many small publishing houses involved in the market. Some of the most significant titles published in English were *Oceanology International; Hydrospace; Ocean Science News; Sea Frontiers; Oceanus;* and *Ocean Industry*. One of the first publications to mix scientific papers with shorter pieces in this style was *Nature* during the 1920 and 1930s: see Melinda Baldwin, "'Keeping in the race': physics, publication speed and national publishing strategies in *Nature*, 1895–1939," *BJHS*, 47:2 (June 2014): 257–280.
47. The White Fish Authority was founded in 1951 as a government body intended to develop the white fish industry. Although it supported the industry in a variety of ways as a trade body, it was primarily intended to be a researching body, so as to improve equipment, and improve catches through fisheries research.
48. This appeared on the journal's editorial page.
49. This attitude was articulated in Brundrett's 'On the Neglected Sea' (1963); Frederick Brundrett, "The Neglected Sea," *Journal of Navigation*, 16:3 (1963):332–342;DOI:http:/dx.doi.org/10.1017/S0373463300020543 (accessed 1/5/14).

50. Jacob D. Hamblin, *Oceanographers and the Cold War: Disciples of Marine Science* (Seattle: University of Washington, 2005): 135.
51. M.J. Tucker, R. Bowers, "Oceanographic Equipment: Present Needs, Future Trends," *Hydrospace,* 1:1 (November, 1967): 48.
52. The complete editorial board was as follows (job titles as they appeared on the title page) and reflected the wide range of interests in applications of science and technology to the exploitation of ocean resources; Rear Admiral Rawson Benett (ret.) [Former Chief of Naval Research (US)], Arthur C. Clarke [British Scientist and Writer], John H Clotworthy [General Manager, Underseas Div., and vice president Westinghouse Defense & Space Center], Capt. Jacques-Yves Cousteau [Director, Oceanographic Museum, Monaco], Dr Carey Croneis [Chancellor, Rice University], Dr Harold E. Edgerton [Professor of electrical measurements MIT], Dr Joel W. Hedgpeth [Director, Marine Science Laboratory, Oregon State University], Edwin A. Link [Chief Marine Consultant and Director, Ocean Systems Inc.], Rear Admiral Manley H. Simons (ret.) [Executive Secretary, Marine Technology Society], Dr F. N. Spiess [Director, Marine Physical Laboratory, Scripps Institution of Oceanography], Dr Athelstan Spilhaus [Dean, Institute of Technology, University of Minnesota], Richard C. Vetter [Executive Secretary, Committee on Oceanography, National Academy of Sciences-National Research Council], Allyn C. Vine [Senior Oceanographer, Woods Hole Oceanographic Institution].
53. Frederick Brundrett, "Where are the engineers?," *Hydrospace: quarterly review of ocean management,* 1:1 (1967): 17–18.
54. Unlike *Oceanology International* fellow journal *Hydrospace* did not publish details of its editorial board and I have been unable to confirm that there ever was one; instead it advertised for submissions to be sent to the editor, Robert Barton, who held a consistent editorial voice in the subsequent issues after Brundrett's first editorial.
55. Barton, *Oceanology Today,* 25.
56. Melbourne Briscoe, "A Brief Survey of Military Oceanography at the NATO SACLANT ASW Research Centre," *US Navy Symposium on Military Oceanography: Proceedings Vol. 1,* US Naval Oceanographic Office (1969) 67; www.biodiversitylibrary.org/ia/usnavysymposiumo669usna (accessed 1/5/14).
57. Sir Anthony Laughton, interview by Paul Merchant, 29 November 2010, British Library, C1379/29, Transcript (Track 7): 152.
58. Gary Kroll, *America's Ocean Wilderness: A Cultural History of Twentieth Century Exploration,* (University of Kansas Press, 2008): 152–188.
59. John Swallow, interview by Margaret Deacon, 30 November 1994, Personal Collection of John Gould, Transcript, 20–4.

60. Barton, *Oceanology Today*, 166–7.
61. Barton, *Oceanology Today*, 168. The construction of a side-scan sonar was much more in line with British financial capabilities, compared with the lavish amounts expended by the USA on the Alvin submersible (US$3 million); see Naomi Oreskes, "A Context of Motivation: US Navy Oceanographic Research and the Discovery of Sea-Floor Hydrothermal Vents," *Social Studies of Science*, 33:5 (2003): 707.
62. Barbara Moran, *The Day we lost the H-Bomb: Cold War, Hot Nukes, and the worst Nuclear Weapons Disasters in History* (Random House, 2009): 130.
63. The US Geological Survey maintains a website on the history of the 1980s GLORIA coastal mapping, http://coastalmap.marine.usgs.gov/gloria/. J.V. Gardner, M.E. Field, D.C. Twichell, *Geology of the United States Seafloor: The View from GLORIA* (Cambridge: Cambridge University Press, 1996).
64. Barton, *Oceanology Today*, 164–7.
65. Anthony Adler, "The Ship as Laboratory: Making Space for Field Science at Sea," *Journal of the History of Biology* (2014): 333–362.
66. Viktor K. McElheny, "Oceanography in Britain," *Science*, 1966. Deacon commonly argued that the Institute had to maximise the value from resources; see Chap. 4 of this book.
67. McElheny, "Oceanography in Britain," *Science*, 1966.
68. Oreskes, "A Context of Motivation," 697–742.
69. Cantoni and Veener, "Underground and Underwater," 55–61.
70. This was discovered to be the cause of the *Sea Gem*'s collapse on 27 December 1965.
71. Jon Agar, "What Happened in the Sixties?," *BJHS*, 41:4 (2008): 567–600.
72. This was specifically embodied by the UN Committee on the Peaceful Uses of the Seabed and the Ocean Floor beyond the Limits of National Jurisdiction, which sat from 1968–69; an overview of the arguments can be found in V.P. Nanda, "Some Legal Questions on the Peaceful Uses of Ocean Space," *Virginia Journal of International Law*, (May, 1969) 343–407.
73. Turchetti, "Sword, Shield and Buoys," 220.
74. Jacob D. Hamblin, *Poison in the Well*, 224.
75. A.C. Simpson, *The Torrey Canyon Disaster and Fisheries* (Ministry of Agriculture, Fisheries and Food; Laboratory Leaflet No. 18, February 1968): 1.
76. Ibid., 33–36.
77. *Torrey Canyon* Meeting 2, 26 March 1967, CAB 130/318, TNA (London).
78. Charles More, *Black Gold: Britain and oil in the Twentieth Century* (London: Continuum, 2009): 127.

79. 'The Spread of the Torrey Canyon Oil', read at the British Association Meeting – Leeds 1968, GERD Papers, D8/20, NOC Library (Southampton).
80. Ibid.
81. Ibid. Deacon pointed out that there was extensive German research produced in the wake of the 8,000 tons of crude which leaked from the *Gerd Maersk* at the mouth of the Elbe in January 1955.
82. Memo 'Projected British Oceanographical Institute', 6 October 1947; J.N. Carruthers, GERD Papers, M3/2/7, NOC Library (Southampton).
83. Op.. cit. no.77.
84. Cabinet Office, *The Torrey Canyon: Report of the Committee of Scientists on the Scientific and Technological Aspects of the Torrey Canyon Disaster* (HMSO, 1967): 47, para.176.
85. Anon., 'Torrey Canyon: A Postscript', *The Lancet* (27 July, 1968): 205; Anon., 'After the Torrey Canyon', *The Lancet,* (6 April 1968): 735; J.E. Smith, *"Torrey Canyon" Pollution and Marine Life* (Cambridge University Press, 1968).
86. Max Blumer, "Oil Pollution of the Ocean," *Oceanus* (1969): 3–7.
87. George R. Hampson and Howard L. Sanders, "Local Oil Spill," *Oceanus* (1969): 8–11.
88. See, 'editorial', *Marine Pollution Bulletin* 1:1 (January, 1970); Rachel Carson, *Silent Spring* (Boston: Houghton Mifflin, 1962); Ruth Harrison, *Animal Machines* (London: Vincent Stuart, 1964).
89. P. Wells, *et al.*, "From mimeos to E-copy – a tribute to Professor R B Clark, founding editor of the Marine Pollution Bulletin," *Marine Pollution Bulletin,* 46:9 (Sept. 2003): 1051–4.
90. Ibid.
91. P. Cloud, "Mined out! Our diminishing mineral resources," *Ecologist,* 1:2 (August, 1970): 25–29.
92. A.J., Puffett, "Prevention of marine pollution," *Ecologist,* 1:7 (January, 1971); J.D. George, "Can the seas survive?," *Ecologist,* 1:9 (March, 1971); R.T. Coon, "Aluminium and Anglesey," *Ecologist,* 1:12 (June, 1971); E. Owen, "The man who sued the Torrey Canyon," *Ecologist,* 1:13 (July, 1971).
93. John Sheail, *An Environmental History of Britain* (Palgrave: London, 2002): 17.
94. Jacob D. Hamblin, "Environmentalism for the Atlantic Alliance: NATO's Experiment with the 'Challenges of Modern Society'," *Environmental History,* 15:1 (2010): 54–75.
95. Ron Doel, "Constituting the post-war earth sciences," *Social Studies of Science,* 33:5 (2003): 635; discussed in Turchetti, "Sword, Shield and Buoys", 227.

96. Letter from H.W.K. 'Bill' Kelly (DNPR, MoD) to Deacon, 26 August 1968, GERD Papers D5/3, NOC Library (Southampton).
97. Letter from Deacon to H.W.K. 'Bill' Kelly (DNPR, MoD), 30 August 1968, GERD Papers, D5/3, NOC Library (Southampton).
98. 'USSR Proposal for co-operative studies of the Southern Ocean', Minutes of the BNCOR Executive Committee, 22 May 1967, authored by G. Deacon, GERD Papers, D4/7, NOC Library (Southampton).

References

Adler, Anthony. "The Ship as Laboratory: Making Space for Field Science at Sea." *Journal of the History of Biology* 47 (2014): 333–362.

Agar, Jon. "What Happened in the Sixties?" *British Journal for the History of Science* 41 (2008): 567–600.

Baldwin, Melinda. "'Keeping in the Race': Physics, Publication Speed and National Publishing Strategies in *Nature*, 1895–1939." *British Journal for the History of Science* 47, no. 2 (June 2014): 257–280.

Bambridge, A. "BP Steps on the Gas." *The Observer.* 26 September 1965.

Barton, Robert. *Oceanology Today.* London: Aldus, 1970.

Blumer, Max. "Oil Pollution of the Ocean." *Oceanus* (1969): 3–7.

Briscoe, Melbourne. "A Brief Survey of Military Oceanography at the NATO SACLANT ASW Research Centre." *US Navy Symposium on Military Oceanography: Proceedings Vol. 1.* Washington, D.C.: US Naval Oceanographic Office, 1969.

Brundrett, Frederick. "The Neglected Sea." *Journal of Navigation* 16 (1963): 332–342.

Brundrett, Frederick. "Where Are the Engineers?." *Hydrospace: Quarterly Review of Ocean Management* 1 (1967): 17–18.

Cantoni, Roberto, Leucha Veneer. "Underground and Underwater: Oil Security in France and Britain During the Cold War." In *The Surveillance Imperative: Geosciences During the Cold War and Beyond.* Edited by Simone Turchetti and Peder Roberts. London: Palgrave, 2014.

Carruthers, James N. "A Plea for Oceanology." *Deep Sea Research* 2 (1955): 247.

Carson, Rachel. *Silent Spring.* Boston: Houghton Mifflin, 1962.

Cloud, P. "Mined Out! Our Diminishing Mineral Resources." *Ecologist* 1, no. 2 (1970): 25–29.

Cockburn, Robert. "Science, Defence and Society." *Journal of the Royal Society of Arts* 115 (1967): 547–559.

Coon, R.T. "Aluminium and Anglesey." *Ecologist* 1, no. 12 (June, 1971): 9–13.

Doel, Ron E. "Constituting the Post-War Earth Sciences." *Social Studies of Science* 33 (2003): 635–666.

Florey, Lord O. M. "Address of the President, Lord Florey, O.M., at the Anniversary Meeting, 30 November 1965." *Proceedings of the Royal Society of London. Series B, Biological Sciences* 163 (1966): 425–434.

Gaber, N.H., D.F. Reynolds. "Economic Opportunities in the Oceans." *Battelle Technical Review* 14 (1965): 5–11.

Gardner, J.V., M.E. Field, D.C. Twichell. *Geology of the United States Seafloor: The View from GLORIA*. Cambridge: Cambridge University Press, 1996.

George, J.D. "Can the Seas Survive?" *Ecologist* 1, no. 9 (March, 1971): 4–9.

Goss, Richard. "British Ports Policies Since 1945." *Journal of Transport Economics and Policy* 32 (1998): 51–71.

Hamblin, Jacob D. *Oceanographers and the Cold War: Disciples of Marine Science*. Seattle: University of Washington, 2005.

Hamblin, Jacob D. "Environmentalism for the Atlantic Alliance: NATO's Experiment with the 'Challenges of Modern Society'." *Environmental History* 15 (2010): 54–75.

Hampson, George R., Howard L. Sanders. "Local Oil Spill." *Oceanus* (1969): 8–11.

Harrison, Ruth. *Animal Machines*. London: Vincent Stuart, 1964.

HMSO. *The Torrey Canyon: Report of the Committee of Scientists on the Scientific and Technological Aspects of the Torrey Canyon Disaster*. London: HMSO, 1967.

Hünemörder, Kai. "Environmental Crisis and Soft Politics: Détente and the Global Environment, 1968–1975." In *Environmental Histories of the Cold War*. Edited by J.R. McNeill and Corinna R. Unger. Cambridge: Cambridge University Press, 2010.

Krige, John. "NATO and the Strengthening of Western Science in the Post-Sputnik Era." *Minerva* 38 (2000): 81–108.

Krige, John. "The Peaceful Atom as Political Weapon, Euratom and American Foreign Policy in the Late 1950s." *Historical Studies in the Natural Sciences* 38 (2008): 5–44.

Kroll, Gary. *America's Ocean Wilderness: A Cultural History of Twentieth Century Exploration*. University of Kansas Press, 2008.

McElheny, Victor K. "Oceanography in Britain: Significant New Support." *Science* 153(1966): 727–8.

Michaelis, A. "Neglected Riches of the Deep." *Daily Telegraph* 22 January 1966.

Moran, Barbara. *The Day We Lost the H-Bomb: Cold War, Hot Nukes, and the Worst Nuclear Weapons Disasters in History*. London: Random House, 2009.

More, Charles. *Black Gold: Britain and Oil in the Twentieth Century*. London: Continuum, 2009.

Nanda, V.P. "Some Legal Questions on the Peaceful Uses of Ocean Space." *Virginia Journal of International Law* 9 (May, 1969): 343–407.

Nehring, Holger. "National Internationalists: Britain and West German Protests Against Nuclear Weapons, the Politics of Transnational Communications and the Social History of the Cold War, 1957–1964." *Contemporary European History* 14 (2005): 559–582.

Oreskes, Naomi. "A Context of Motivation: US Navy Oceanographic Research and the Discovery of Sea-Floor Hydrothermal Vents." *Social Studies of Science* 33 (2003): 697–742.

Owen, E. "The Man Who Sued the Torrey Canyon." *Ecologist* 1 (1971): 22–23.

Puffett, A.J. "Prevention of Marine Pollution." *Ecologist* 1, no. 7 (January, 1971): 24–25.

Roberts, Peder. "'What Has All This Got to Do with Science?' The Rhetoric of Scientific Devotion in British Government Plans for the International Geophysical Year." *Proceedings of the 3rd Workshop of the SCAR Action Group on the History of Antarctic Research.* Url: https://kb.osu.edu/dspace/bitstream/handle/1811/BPRC_Tech_Rept_2011-01.pdf (accessed: 19/2/2014).

Rose, Hilary, Stephen Rose. *Science and Society.* London: Allen Lane, 1969.

Sheail, John. *An Environmental History of Britain.* London: Palgrave, 2002.

Simpson, A.C. *The Torrey Canyon Disaster and Fisheries.* Ministry of Agriculture, Fisheries and Food; Laboratory Leaflet No. 18, February 1968.

Smith, J.E. *"Torrey Canyon" Pollution and Marine Life.* Cambridge: Cambridge University Press, 1968.

Solovey, Mark. "Project Camelot and the 1960s Epistemological Revolution: Rethinking the Politics-Patronage-Social Science Nexus." *Social Studies of Science* 31 (2001): 171–206.

Stevenson, Iain. *Book Makers: British Publishing in the Twentieth Century.* London: British Library, 2010.

The Guardian. "Research Reshaped: Commons Debate on Technology." 12 December 1964.

The Guardian. "Six Firms Sponsor Sea Study." 17 July 1965.

The Lancet. "Torrey Canyon: A Postscript." 27 July 1968.

The Lancet. "After the Torrey Canyon." 6 April 1968.

Tucker, M.J., R. Bowers. "Oceanographic Equipment: Present Needs, Future Trends." *Hydrospace: Quarterly Review of Ocean Management* 1 (1967): 48.

Tucker, Richard P. "The International Environmental Movement and the Cold War." In *The Oxford Handbook of the Cold War.* Edited by Richard H. Immerman and Petra Goedde. Oxford: Oxford University Press, 2013.

Turchetti, Simone, Simon Naylor, Katrina Dean, Martin Siegert. "On Thick Ice: Scientific Internationalism and Antarctic affairs, 1957–1980." *History & Technology* 24 (2008): 351–376.

Turchetti, Simone. "Sword, Shield and Buoys: A History of the NATO Sub-Committee on Oceanographic Research, 1959–1973." *Centaurus* 54 (2012): 205–231.

Wells, P., et al. "From Mimeos to E-copy – A Tribute to Professor R B Clark, Founding Editor of the Marine Pollution Bulletin." *Marine Pollution Bulletin* 46 (2003): 1051–4.

CHAPTER 8

Epilogue: The Retirement of George Deacon

Deacon's retirement in 1971 was a turning point in the history of British oceanography. The management of various oceanographic research institutions in Britain came under a new director with responsibility for a new institutional structure named the Institute of Oceanographic Science. Since the new institute included a network of laboratories scattered around the country, rather than just the centre in Wormley, the restructuring resulted in the new director playing a completely different role, concerned more with the day-to-day running of the new network of laboratories than lobbying in Whitehall. Similarly, the oceanographic programme itself shifted, turning further away from commitment to military oceanography projects and towards environmentalism and environmental studies.

The scientific and political stage upon which ocean scientists in Britain conducted their science on the ocean frontier was fundamentally changed when George Deacon was finally forced to retire as the NIO director.[1] Finding a replacement for a figure who had been a relatively junior scientist before taking over the directorship, but who had now become a recognisable international oceanographic expert, proved challenging.[2] Deacon had to accept that his successor was going to be another manager heading a different sort of research institute. Unsurprisingly, the NERC committee appointed to search for Deacon's successor approached the Cambridge geophysicist Edward Bullard, to whom it was now clear that

S. A. Robinson, *Ocean Science and the British Cold War State*,
Palgrave Studies in the History of Science and Technology,
https://doi.org/10.1007/978-3-319-73096-7_8

any new incumbent director would find it difficult to come out of the shadow of Deacon's leadership.[3]

Suitable candidates were limited. Initially Bullard approached Anthony Laughton whilst they were both attending the Joint Oceanographic Assembly conference "The Ocean World" in Tokyo (Japan) in September 1970.[4] In correspondence following the conference with Raymond Beverton (the NERC Secretary), it became apparent that Laughton was extremely surprised at being approached, and was not sure that he was ready to—as he saw it—put his science to one side in order to deal with the bureaucracy the position of director entailed.[5] Bullard was surprised by the apprehension; he knew Laughton because he had been his PhD supervisor, and Laughton's work on sea-floor geophysics was connected with similar work being undertaken within the Department of Geodesy and Geophysics at Cambridge. Eventually, Bullard approached other senior scientists at the NIO and discovered that they were disinclined to consider even applying for the position. He had to admit to Beverton that it appeared that despite Deacon's rhetoric about the need for an expansion of marine science, he had not actually "groomed" a suitable or even a willing successor at Wormley. Bullard wrote to Beverton:

> It is a bit odd that we haven't got an obvious candidate considering all the money we have put into oceanography over the last twenty years. Maybe NIO has been kept too isolated from the people who use oceanography and its staff has not had a chance to develop the width of interest and the connections that we want in a director. I think we should deliberately encourage relations between NIO and the Navy, the oil companies, and marine engineering firms.[6]

Bullard decided to go and speak to the senior scientists and see if they had a suggestion. The name that arose from these discussions was, unexpectedly, Henry Charnock.

Charnock was not a favourite of Deacon. Following his attempt to unionise scientists at the Institute after the NERC establishment in 1965, Charnock had left the NIO at Deacon's behest and had in 1966 established a department of oceanography at the University of Southampton.[7] He was not the only member of staff to leave; 46 from a total of 200 staff left during the first two years of NERC control.[8] Charnock was a marine chemist/physicist and since the Second World War he had been involved in the study of air–sea interaction.[9] During the early history of the NIO he had acted as Deacon's assistant, working on military-scientific physical oceanographic programmes (as seen in Chap. 4).[10] Later, he was employed at the NATO SACLANT Antisubmarine Warfare Centre (SACLANTCEN) in La Spezia, Italy.[11] Like Laughton he was almost the perfect composite in terms of research interests and experience, having spent time both in military and university environments, and being a specialist in the physical rather than the biological side of oceanography without being just another geophysicist. However, Deacon felt betrayed by Charnock's attempt to establish a trade union at the NIO, especially since this happened during the period of uncertainty that followed the Trend Report (1963).

In order to avoid opening old wounds Deacon agreed with Bullard to leave the Institute for a while, supported by a fellowship from the NERC. In a personal exchange of letters between the two it appears that both men now realised that the wartime oceanographers were moving on—a generational shift was occurring.[12] This transition was not a coup, or a revolution, it was an orderly transfer of power. However, the events serve as another example of Deacon's weakened position. The former director was not to be around when his own successor was first appointed. After that Deacon went to sea for the first time in years, leaving the new director with a large institute to manage without the direct support and supervision of his predecessor.

In 1949 Deacon had been given the reins of the NIO with only a small staff; it was now a large research centre with well-developed research groups. Charnock was soon challenged as director; in the early 1970s the new Institute of Oceanographic Sciences (IOS) had to cope with a reduced budget, since the NERC, in order to meet the challenge of an impending economic recession, agreed to combine several smaller marine institutes under the IOS banner. Charnock was thus made director of four UK oceanographic institutions (NIO: Wormley, Tidal Institute: Liverpool, Coastal Sedimentation Unit: Taunton, Research Vessel Services: Barry, South Wales), driving between each on different days of the week. Ultimately Charnock had been given the power that Deacon had always aspired to. The new director was responsible for almost all of the non-university civilian oceanographic research in the United Kingdom. It was, however, a poisoned chalice, given the increased number of managerial tasks that Charnock had to deal with and his reduced influence in Whitehall.[13] The role of director had, due in no small part to the change of personality, shifted from being the leader of ocean science nationally and British oceanography internationally as Deacon had seen himself, to more of the scientific management that was becoming increasingly common in industrial research. When Deacon retired, oceanography was therefore no longer the same discipline that he had learned to administer in the previous twenty years, and—at least as far as Britain was concerned—would probably never again be so closely tied to military fortunes.

Notes

1. Under the rules of the Royal Navy Deacon would have reached retirement age in 1965; moving to the NERC meant that Deacon could continue until he was 65, the civil service retirement age.
2. Letter from Edward Bullard to Professor Wynne-Edwards (Chairman, NERC), 15 July 1970, ECBP 100.4.84/F.37, CAC.
3. Letter from Professor Wynne-Edwards to Bullard, 7 July 1970, ECBP, 100.4.84/F.37, CAC.
4. Letter from Anthony Laughton (SSO at NIO) to Bullard, 25 October 1970, ECBP, 100.4.84/F.37, CAC.
5. In particular letter from Bullard to Raymond Beverton (Secretary, NERC), 4 December 1970, ECBP, 100.4.84/F.37, CAC.
6. Ibid.
7. D.E. Cartwright, "Henry Charnock, C.B.E. 25 December 1920–27 November 1997: Elected F.R.S. 1976," *Biogr. Mems Fell. R. Soc.* (1999): 42.
8. Henry Charnock, "George Edward Raven Deacon. 21 March 1906–16 November 1984," *Biographical Memoirs of Fellows of the Royal Society,* 31 (1985): 133.
9. D.E. Cartwright, "Henry Charnock," *Biogr. Mems Fell. R. Soc.* (1999).
10. See Chap. 4, 117–119.
11. Having left the NIO to take up a readership at Imperial College, Charnock was reinstated on the basis that he take up a three-year posting to La Spezia in 1958; Cartwright, "Henry Charnock," 41.
12. Letter from Deacon to Bullard, 9 March 1971; letter from Bullard to Deacon, 12 March 1971; letter Deacon to Bullard, 18 March, ECBP, 100.4.84/F.37, CAC.
13. Cartwright, "Henry Charnock," 43–44.

CHAPTER 9

Conclusion: Situating Britain and the Sea in the Cold War

The primary objective of this book has been to provide a history of how oceanographic science in Britain during the Cold War interacted with its state patrons. Three themes have emerged which have driven its analytical framing. As outlined in the introduction, these themes were prompted by questions raised by recent scholarship on the contemporary history of science in general, and the Cold War in particular. First, this work set out to undertake a study of the management of ocean science, following Steven Shapin's suggestion that management studies are a suitable alternative to histories of individuals, or collective scientific histories of the teamwork of modern science.[1] Second, the connections between ocean science and surveillance were explored, and the ways in which oceanographic research shaped (and was shaped by) the geopolitical and geostrategic understanding of the ocean environment in the Cold War were considered. Finally, this book has striven to break down the monolithic understanding of government research funding, revealing the inner workings of Whitehall's policy structures and accounting for oceanography's growth as a scientific discipline through the acquisition of precious government resources (inter-departmental, intra-admiralty, and scientific). The director of the National Institute of Oceanography, George Deacon, is the lens through which I have looked at each theme.

Through Deacon's career one can learn a great deal more about the establishment and growth of military oceanography as central pursuit of

S. A. Robinson, *Ocean Science and the British Cold War State*,
Palgrave Studies in the History of Science and Technology,
https://doi.org/10.1007/978-3-319-73096-7_9

the NIO in the early years of its existence. Sitting at the centre of important policy networks in Britain, Deacon was the conduit through which oceanographic studies in Britain undertook a defence research agenda distinctive to the Cold War era; he was responsive to the surveillance needs of the Cold War; and he directed the resources necessary for these commitments. Just as Deacon's career has been such a central focus of this book, so his retirement provides an appropriate segue into a summary of what each of these themes tells us about oceanographic sciences in Cold War Britain.

Oceanography in the Cold War has been described here as a disruptive force which aimed to stake a claim to the very authority of knowledge of the sea that had lain for centuries with naval rather than scientific power. Whilst this may be true of many nations in which oceanography was a rather underdeveloped discipline before 1945, in Britain oceanography during the Cold War era can be better defined as the struggle between tradition and innovation. It is too easy to define tradition in terms of institutions, but in reality tradition versus innovation more often than not came down to individuals. The innovation of creating the NIO came from a Hydrographer (Edgell) who was known for pushing forward his service. The repossession of military oceanography by a different personality (Irving) undertaking the same role during the 1960s was debated on the premise of traditional roles and the separation of military and civilian spheres. Deacon liked to present himself as the innovator of oceanography, pushing back the constraints of tradition. This can most consistently be seen through his advocating of an equitable hybrid approach to ocean science (seeing biology-physics-chemistry as mutually beneficial to any study), rather than a focus on fisheries research that was seen as of economic benefit to the nation but of lesser urgency in the Cold War climate. As the power of Deacon's network was constructed solely on the basis of other actors (such as the Hydrographer), it was *where* Deacon stood on issues, or rather *who* he stood alongside, that was critical to the maintenance of power within the network.

As has been shown, the "national" was very important within the NIO's title for many years. Deacon was careful to maintain this primacy and centralisation, for it increased the power of both his network and the institution. That is not to say that he successfully monopolised research resources, although that was a claim made by his detractors, but he certainly attempted to do just that. In lobbying for the NIO, Deacon was by extension putting himself at the centre of the oceanographic community.

This required skillful handling of peers and colleagues so as to maintain a network which revolved around him as the central mediating actor with those in other networks within the Admiralty and Whitehall. The intention is not to demonise Deacon here; there is a risk of presenting him as acting only for himself and in his own self-interest. Often he was chosen, by those in the Admiralty or wider civil service, as their "safe pair of hands" prepared if necessary to undertake "dirtier" tasks.

This has illuminated the overall analytical narrative of this book. Often Brundrett and Deacon worked as a "double act", depending on whether science or government needed persuading of the national-security implications of oceanographic research. Others came and went over the years. From the field of oceanography they were James Carruthers and Edward Bullard, although Bullard along with the American oceanographer Roger Revelle ultimately turned against Deacon's supremacy in the British arena. In the Admiralty Deacon was largely supported by the incumbent Hydrographer (Edgell) during the war and by the chief scientific adviser Frederick Brundrett into the 1950s, but undermined by the Hydrographer (Irving) during the 1960s. He also found allies amongst wartime scientific administrators who rose to higher office after the conflict, Charles S. Wright (the post-war CRNSS) being chief amongst them. Figures within the civil service, and outside the military, became important over time as ocean science drifted away from under military control and received acclaim along with the other geosciences which had co-created their own hero narratives of science at geographical extremes, whether in space, on ice, or at sea. This changing cast of supporting actors makes the longevity of Deacon's role not only surprising but also critical in any attempt to explain the course of oceanographic affairs at the height of the Cold War. His use of his international position to shape events at home recasts notions of the national, international, and transnational. This international dimension is important in understanding how Deacon used events such as the IGY and bodies such as NATO as a means to gain power at home.

Deacon is a much more representative figure as the scientific administrator-manager than the celebrities of post-war science who in the best traditions of hagiographical histories receive so much attention. In the same way as Charles Thorpe uses Oppenheimer as an exemplar of his age, so too I believe that Deacon provides an example of the scientific administrator so indicative of post-war science, but seldom seen in historical treatments of British Cold War science, and in this way he provides a more representative object of study.[2] Deacon's position, just below that of the recognisable "leaders" of science, meant that he was not tied to any particular science policy perspective; rather he had to work with the establishment and institutional landscape before him. This did not mean that he was an idle bystander; instead he often tried to influence science policy, rather than having the ability to shape it himself, through his access to actors within other networks. In this way he was a key mediator for oceanography in the political-scientific milieu, not just at home but also internationally. With this in mind it is better to see Deacon as a navigator of events and circumstances. Of the many terms we may use for this—middleman, pragmatist, mediator, or even translator—all are partially applicable.

The correlation between the work of ocean scientists in Britain during the early Cold War and the work of other senior ocean scientists who worked to secure funding to expand their counterparts' scientific programme can only be explained through a close study of the interface between politics and science, that is, a study of the bureaucratic machine. As has been demonstrated, the systems, rules, and procedures all devolve their power to the decisions made by individuals to establish the terms of policy outcomes. Here the recent conclusions of principal–agent theorists are borne out, namely that both principals (military, government, Treasury officials) and agents (ocean scientists and scientific bureaucrats, administrators, managers) seek to acquire either as much return for their investment, or alternatively as many resources as possible. As a result each lobbies the other. To this end Deacon "spun" research to attract funding, in an attempt to persuade those in other networks that he could deliver policy-relevant research without strict monitoring. Here this notion has been challenged by the suggestion that Deacon was not an omniscient actor who always knew what was wanted of oceanography. Rather he might have guessed, or at other times tried to inform his patrons of the kind of science they needed, and often the type of science he and his scientists wanted to perform.

This is where Deacon's role as a mediator between networks is cast into the sharpest relief. Here the continuing significance of connected actors whose careers developed in parallel is central. Deacon was never rated a great scientist by his peers, but he was valued as an administrator. As an administrator he knew that as a scientist he often had to deliver the expectations of others without necessarily agreeing with them. This need to do the devil's work to reap the financial rewards was a consistent occurrence throughout his career. As shown in Chap. 2, he had to be seen to conduct

British trials on the use of the American bathythermograph in ASW, but knowing the quality of the research, he wrote his report before British trials had even been completed. Later, during the 1950s, he had to deploy NIO scientists to undertake trials in the Mediterranean that he knew would fail, but withdrawal would have angered his military backers, putting future financial patronage at risk (Chap. 4). During the 1960s he had to find a way of working within a shifting scheme of expectations for military oceanography, whilst maintaining the NIO and contributing to civilian and military science (Chap. 6). This was achieved through a network of sympathisers, or perhaps more colloquially friends, for whom Deacon could return favours, thus essentially helping to realise their vision.

Throughout the period of Deacon's leadership of the NIO the rivalry between physical and biological oceanography regularly resurfaced. At the foundation of the NIO Deacon managed to appease both the biologists and the military officers through a promise to ensure that even under Admiralty patronage a biological programme would be maintained, and on the other hand that any biological programme would complement the military-physical research. Maintaining the impression that this was a sincere promise became increasingly difficult because of the growth through international military bodies such as NATO of sponsored physical oceanographic research. As civilian international oceanographic bodies also became increasingly populated with physical oceanographers, the initial promise seemed to have been eroding because of circumstances that appeared to lie at Deacon's door, as the Navy's man. Eventually leading figures of the biological oceanography community such as Neil Mackintosh attempted to move against Deacon by withdrawing the whale research element from the NIO and returning it to the Natural History Museum in London. As soon as Deacon was not the head of what appeared to be a cohesive scientific community his position as a spokesman for British oceanography was undermined. And, as outlined in the introduction, those making policy in government prefer dealing with identifiable leaders of interest groups as it streamlines the policy process in terms of time, money, and effort on the part of administrators.

Ultimately Deacon's hegemonic position at the head of British oceanography was undermined by a new generation of military officers and bureaucrats rather than scientists. Although his prestige helped him maintain his position as director, within the scientific community around the Royal Society, as the British oceanographic representative at NATO, and within wider international oceanography, his non-scientific position was dependent on shifting policy ambitions and directives. The strong tradition of military science within the British military state (or as David Edgerton has called it, the British "warfare" state[3]) was strengthened by a reliance on special allies in the scientific community working on defence operations and as a consequence could reap a benefit from science in terms of increased security. However, as these "special allies" were slowly replaced by new figures who had not undertaken military science under wartime conditions and had less affinity and connection with the military world, and as the political notion of science as a social resource arose, this military-scientific state changed.

Science became the beacon upon which Britain's economic problems could be eradicated, through modernity brought about by nationalised industry benefiting from nationally funded scientific research. As John Cockcroft put it, science could be used for the benefit of mankind rather than being put towards its destruction. By the late 1960s the militarisation of civilian science was increasingly viewed as problematic.[4] This changing political landscape created new levels of scientific governance that were not unique to the oceanographic sciences; it prompted a turn to a new generation of scientists and a turn away, at least in public, from scientists who seemed to belong to an earlier era. Whereas Deacon had been active in shaping his own identity earlier in his career, he was less able to do so by his retirement. By the time he stepped down as NIO director his role had changed dramatically. No longer did he influence policy from within the corridors of power: he was marginalised and with friends and patrons already having retired or moved on, his network was much weaker than it had once been.

From this turbulence, however, there emerged a new framing for the oceanographer: that of the scientific intelligence gatherer. This further dimension of Deacon's career, spying on science, came at the height of international oceanographic cooperation. For example, Deacon used his privileged position on the IGY committees to gather and judge available open source on Soviet oceanographic capabilities and intentions for the Directorate of Scientific Intelligence (Admiralty) which reported straight into Britain's central intelligence committee, the JIC. As shown in the second half of this book, by the end of the 1950s the idea of oceanographic science being used to assist in the deployment of naval forces to control the ocean environment took a technological step forward with the concept of fixed surveillance systems constantly monitoring the ocean (Chap. 5). The gathering of "human" intelligence was complemented by the instrumental intelligence of the oceans being assembled by oceanographers, reaffirming the importance of gaining information on enemy capabilities at sea by exploiting both scientists like Deacon's international positions and the devices that their work contributed to generate. That this strategic reality developed in tandem with geoscientists' desire for synoptic measurement of the earth was no coincidence. Science and the military desired the same extensive knowledge of the earth; it was just that their intended application of the knowledge appeared different.

But was it really that different? Chapter 5 demonstrated that surveillance and monitoring and its duplicity of objectives were used by Deacon and others as a justification for the close cooperation between military and civilian scientists in the ocean. The fact that ultimately (Chap. 7) oceanographers should turn to their knowledge of monitoring and surveillance technologies as a means to commercialise their research after 1965, is perhaps the real irony of their separation, but in hindsight wholly inevitable given their previous repackaging of ocean science for military patrons.

This book has attempted to follow Rozwadowski's challenge by keeping a constant focus on the geopolitical outcomes and consequences of oceanographic programmes in order to bring the oceans into the historical field of view.[5] It is argued here that the link between oceanography and geopolitics was shaped by the interconnection between ocean science and technologies/regimes of surveillance and monitoring. Whilst many have revealed the connections between the geosciences and surveillance during the Cold War, this has seldom been done in respect of the oceans.[6] Global surveillance in the Cold War is most often seen as satellites and spy flights; here it is hidden and silent. As this book and the wider project of which it has been a part has shown, the geosciences, through the medium of surveillance and monitoring, played important roles in the furthering of specifically Cold War geopolitical objectives.[7] For oceanographic sciences this took the form of supporting the implementation of fixed surveillance platforms. Whether or not most oceanographers were aware that their research was being used in this way, or until the USS *Pueblo* incident whether many knew that the term oceanography was being used as a cover for ocean espionage, remains difficult to assess.

However, intelligence, surveillance, and monitoring are themes that have run throughout this book. From the very first entanglement of naval science and civilian science in the early months of the Second World War, through to the establishment of RAF Brawdy in 1974, military-scientific relationships were framed around the ability of the scientists to provide expert knowledge of the environmental medium which the Navy sought to control: the ocean. During the war this ocean knowledge was deployed to fight submarines, protect convoys, and destroy enemy mines (Chap. 2). In this process ocean science moved from basic science to applied science to technologies; in essence oceanography took on the sheen of being a 'technoscience'.[8] In the immediate aftermath of the War Deacon "sold" to the Admiralty and ultimately—through intermediaries—to the Treasury the military potential of oceanographic technoscience (Chap. 3), resulting in the proliferation of instruments for oceanographic research deriving from surveillance-related work (Chap. 5). During the 1950s (Chap. 4) the dynamic of civilian oceanography as both a peaceful and a military endeavour, and the difficulties of maintaining resource balance between the sub-disciplines of biological, chemical, and military oceanography, brought tensions at the NIO. Nevertheless, these alternative dimensions of scientific work were fundamental to the longevity and shape of scientific-military-political networks.

To some extent knowledge about the environment brought ocean science into larger discussions about geopolitics, international diplomacy, and global power politics beyond the realm of science. Although these connections are well known and have been identified by recent histories, an explanation of how knowledge flowed from oceanographers to military officers to politicians has been previously neglected. This has been in part explained in this work through the case study of ocean surveillance technologies discussed in Chap. 5. What this study also showed is that even between strong allies there was disagreement as to how the ocean was to be surveyed. Tentative conclusions as to why Britain was more conservative than the United States might be the lack of comparable research effort, in purely monetary terms; however, this provides an unsatisfactory explanation for British efforts. A firmer conclusion may be drawn from the notion of competing conceptions of surveillance. The United States was looking towards building systems of global surveillance, whereas a Britain rapidly losing its empire was forced to reorient its outlook towards local threats to domestic security rather than taking the panoptic approach of their superpower rivals.

This situation of constant compromise has proved an interesting object of study, particularly as regards the relations between scientists, civil servants, and military officers. It suggests that power was shared. As Foucault would have it, power is transient and shifting, rather than rigidly applied as in totalitarian systems, or in democracies at war.[9] Additionally, conceptualising Cold War science policy as a system of compromise differentiates post-war experiences of science from the experience of science during the Second World War. Two significant differences between wartime urgencies and peacetime circumstances can be seen throughout negotiations between Deacon and Whitehall networks: time and money. The use of the Cold War as a metaphor in these negotiations could expand or contract either resource. It is easy to oversimplify different networks attitudes' to these resources. We may believe that scientists seek lots of time and vast sums of money, whereas civil servants desire science quickly and cheaply. By deconstructing the state, this book has shown that these dichotomies are artificial. In the bureaucratic state, attitudes are much more predicated on where someone sits than on where they stand. So whilst officials in the Treasury, who hold the purse strings, were naturally concerned about money, they cared less for time because waiting costs little; for example, the Treasury was willing to allow the scheme to establish an institute to drift until somebody in the Admiralty agreed to pay for it (Chap. 3). On the other hand, the military cared less for money, but were always concerned about staying ahead of their enemy's capabilities. In this situation time was paramount, and cost was a side-effect, as we saw when the governance of the NIO was disputed during the early 1960s and the Admiralty were quite happy to rid themselves of the Institute and bring oceanographic research "in house" where they had greater control (Chap. 6). Deacon and his allies had to negotiate these positions, to work the bureaucratic system, to confront and shape compromise. In this way Cold War-era oceanography in Britain was as much shaped by men in ships as men in suits. Ultimately, we must know as much about the few paper-shufflers as the scientists in ships, institutions, and laboratories to understand the trajectory and construction of ocean science in action.[10]

Notes

1. Steven Shapin, *The Scientific Life: A Moral History of a Late Modern Vocation* (Chicago: University of Chicago Press, 2008).
2. Charles Thorpe, *Oppenheimer: The Tragic Intellect* (Chicago: University of Chicago Press, 2006).
3. David Edgerton, *Warfare State: Britain 1920–1970* (Cambridge: Cambridge University Press, 2005).
4. Whilst this began with the anti-nuclear movement, it came to encompass all science that was linked to nuclear weapons; oceanographers' support of

ASW and the development of surveillance techniques implicated the military aspects of the NIO just as much as nuclear scientists elsewhere. Helge Kragh, *Quantum Generations: A History of Physics in the Twentieth Century* (Princeton, NJ: Princeton University Press, 1999): 394–404.

5. Helen Rozwadowski, "Introduction," *Isis*, 105:2 (2014): 335–337.
6. See for example, Simone Turchetti and Peder Roberts, *The Surveillance Imperative: The Geosciences and the Cold War* (Basingstoke: Palgrave, 2014).
7. For outputs from the wider project, see, ibid.
8. Here I want to invoke Latour's remarks 'Technoscience is part of a war machine and should be studied as such'. Bruno Latour, *Science in Action: How to Follow Scientists and Engineers through Society* (Cambridge, MA: Harvard University Press, 1987): 172; Dominique Pestre, "Challenges for the Democratic Management of Technoscience: Governance, Participation and the Political Today," *Science as Culture*, 17:2 (2008): 101–119; Dominique Pestre, (ed.), *Le Gouvernement des technosciences: Gouverner le progrès et ses dégâts depuis 1945* (Paris: La Découverte, 2014); Chris Sneddon, "The 'Sinew of Development': Cold War Geopolitics, Technical Expertise, and Water Resource Development in Southeast Asia, 1954–1975," *Social Studies of Science*, 42:4 (2012): 564–590.
9. Michel Foucault, *Discipline and Punish: the Birth of the Prison* (London: Penguin, 1977 [1991]): 24–31; also see Roger Jervis, "Cooperation under the Security Dilemma," *World Politics*, 30:2 (1978): 167–214.
10. Here I am paraphrasing the closing remarks of Latour, *Science in Action*, 254–257.

Note on Archival Resources

TNA (London): The National Archives, UK, Kew Gardens, London

ADM (Records of the Admiralty, Naval Forces, and related bodies)
BJ (Records of the Meteorological Office)
CAB (Records of the Cabinet Office)
CO (Records of the Colonial Office, Commonwealth and Foreign Offices, and related bodies)
DEFE (Records of the Ministry of Defence)
FO (Records of the Foreign Office)
PREM (Records of the Prime Minister's Office)
T (Records created or inherited by HM Treasury)

NOC (Southampton): National Oceanography Centre Archives, Southampton

GERD Papers (personal papers of former director of the National Institute of Oceanography, George Edward Raven Deacon)
JN Carruthers Papers (personal papers of James Norman Carruthers)

S. A. Robinson, *Ocean Science and the British Cold War State*,
Palgrave Studies in the History of Science and Technology,
https://doi.org/10.1007/978-3-319-73096-7

CAC (Cambridge): Churchill College Archive Centre, University of Cambridge

ECB Papers (Edward Crisp Bullard papers)
BRUN Papers (Sir Frederick Brundrett papers)

Other Archival Collections

Columbus O'Donnell Iselin papers, 1904–1971. Data Library and Archives, Woods Hole Oceanographic Institution. – provided by Peder Roberts

Oral History Interviews

British Library: Oral History of Science

Anthony Laughton, interviewed by Paul Marshal, *British Oral History of Science Interview,* C1379/29 Track 5, p. 106 of Transcript, http://sounds.bl.uk/related-content/TRANSCRIPTS/021TC1379X0029XX-0000A0.pdf

Privately Held Interviews

John Swallow, interviewed by Margaret Deacon 30 November 1994, transcript held by the Royal Meteorological Society (supplied to this author by John Gould, NOC Southampton)

Index

S. A. Robinson, *Ocean Science and the British Cold War State*,
Palgrave Studies in the History of Science and Technology,
https://doi.org/10.1007/978-3-319-73096-7

R

S

The manufacturer's authorised representative in the EU is Springer Nature Customer Service Centre GmbH, Europaplatz 3, 69115 Heidelberg, Germany. If you have any concerns regarding our products, please contact ProductSafety@springernature.com

Printed and bound by CPI Group (UK) Ltd, Croydon, CR0 4YY
05/10/2026
02228128-0006